Cambridge Lower Secondary

Maths

STAGE 8: WORKBOOK

Alastair Duncombe, Belle Cottingham, Rob Ellis, Amanda George, Brian Speed

Series Editor: Alastair Duncombe

Collins

William Collins' dream of knowledge for all began with the publication of his first book in 1819. A self-educated mill worker, he not only enriched millions of lives, but also founded a flourishing publishing house. Today, staying true to this spirit, Collins books are packed with inspiration, innovation and practical expertise. They place you at the centre of a world of possibility and give you exactly what you need to explore it.

Collins. Freedom to teach.

Published by Collins
An imprint of HarperCollinsPublishers
The News Building
1 London Bridge Street
London
SE1 9GF

HarperCollins*Publishers*
Macken House,
39/40 Mayor Street Upper,
Dublin 1, DO1 C9W8,
Ireland

Browse the complete Collins catalogue at
www.collins.co.uk

MIX
Paper | Supporting
responsible forestry
FSC
www.fsc.org
FSC™ C007454

This book contains FSC™ certified paper and other controlled sources to ensure responsible forest management.

For more information visit:
www.harpercollins.co.uk/green

British Library Cataloguing in Publication Data
A catalogue record for this publication is available from the British Library.
Authors: Alastair Duncombe, Belle Cottingham, Rob Ellis, Amanda George, Brian Speed
Series editors: Alastair Duncombe
Publisher: Elaine Higgleton
In-house project editors: Jennifer Hall and Caroline Green
Project manager: Wendy Alderton
Development editors: Phil Gallagher and Jess White
Copyeditor: Laurice Suess
Proofreader: Tim Jackson
Answer checker: Tim Jackson and Jouve India Private Limited
Cover designer: Ken Vail Graphic Design and Gordon MacGlip
Cover illustrator: Ann Paganuzzi
Typesetter: Jouve India Private Limited
Production controller: Lyndsey Rogers
Printed and bound by: Martins the Printers

Acknowledgements

The publishers gratefully acknowledge the permission granted to reproduce the copyright material in this book. Every effort has been made to trace copyright holders and to obtain their permission for the use of copyright material. The publishers will gladly receive any information enabling them to rectify any error or omission at the first opportunity.

p. 119 WHO World Health Statistics made available under CC BY-NC-SA 3.0 IGO

Cambridge International copyright material in this publication is reproduced under licence and remains the intellectual property of Cambridge Assessment International Education.

Third-party websites and resources referred to in this publication have not been endorsed by Cambridge Assessment International Education.

With thanks to the following teachers and schools for reviewing materials in development: Samitava Mukherjee and Debjani Sen, Calcutta International School; Hawar International School; Adrienne Leisztinger, International School of Budapest; Sujatha Raghavan, Manthan International School; Podar International School; Taman Rama Intercultural School; Rakhi Mukerjee, Utpal Sanghvi International School.

Contents

How to use this book

This Workbook accompanies the Collins Lower Secondary Maths Stage 8 Student's Book and covers the Cambridge Lower Secondary Mathematics curriculum framework (0862). This Workbook can be used in the classroom or as homework. Answers are provided in the Teacher's Guide.

Every chapter has these helpful features:

- 'Summary of key points': to remind you of the mathematical concepts from the corresponding section in the Student's Book.

- Exercises: to give you further practice at answering questions on each topic covered in the Student's Book. The questions at the end of each exercise will be harder to stretch you.

- 'Thinking and working mathematically' questions (marked as   ): to help you develop your

  mathematical thinking. The questions will often be more open-ended in nature.

- 'Think about' questions: encourage you to think deeply and problem solve.

1 Negative numbers, indices and roots

You will practice how to:

- Estimate, multiply and divide integers, recognising generalisations.
- Recognise squares of negative and positive numbers, and corresponding square roots.
- Recognise positive and negative cube numbers, and the corresponding cube roots.
- Use positive and zero indices, and the index laws for multiplication and division.

. . . .

1.1 Multiplying and dividing positive and negative integers

Summary of key points

Rules for multiplying and dividing integers

| + × + = + | + × − = − | + ÷ + = + | + ÷ − = − |

| − × + = − | − × − = + | − ÷ + = − | − ÷ − = + |

Examples: $-12 \div -3 = 4$ $\qquad$ $5 \times -2 = -10$ $\qquad$ $-3 \times 4 = -12$

Exercise 1

1 Draw a line to match each calculation with the correct answer.

−3 × −6	−12
36 ÷ −3	12
−36 ÷ 2	18
−48 ÷ −4	−18

2 Tick the true statements and cross the false statements.

a) $6 \times (-8) = -48$ b) $(-20) \div (-4) = -5$ c) $(-42) \div 7 = -6$

d) $(-3) \times (-9) = 27$ e) $(-4) \times 2 \times (-3) = -24$ f) $(-3) - 9 = (-2) \times (-6)$

g) $(-7) - (-10) = (-15) \div (-5)$ h) $45 \div (-9) = (-20) \div 4$ i) $(-3) \times (-3) = (-15) + 6$

3 Work out the result of each calculation. Draw a ring around the odd one out in each set.

a) $(-6) \times (-5)$ $(-90) \div (-3)$ $(-3) \times 10$

b) $(-72) \div 8$ $(-63) \div (-7)$ $(-3) \times 3$

c) $(-44) \div (-4)$ $(-9) - 2$ $(-20) - (-9)$

4 Work out the missing numbers.

a) .... $\times (-5) = 40$ b) $(-52) \div$.......... $= 4$ c) $(-6) \times$.......... $= -42$

d) .... $\div (-9) = 7$ e) $98 \div$.......... $= -2$ f) .... $\div 14 = -3$

5 For each description below, write a calculation with the answer 8.

Include at least one negative integer in each calculation.

a) a product of two integers ....

b) a product of three integers ....

c) a product of four integers ....

d) a quotient of two integers ....

e) a calculation that includes multiplication and division. ....

6 Join the parts to make two correct sentences.

| $(-4)^2$ means | $-(4 \times 4)$ | which equals -16. |

| -4^2 means | $(-4) \times (-4)$ | which equals 16. |

Think about

7 Write a negative number in each box to make a true statement.

$$\boxed{} \times \boxed{} = \boxed{} - \boxed{}$$

Try to find ten different solutions.

Summary of key points

A positive number has …	A negative number has …
• two square roots, one positive and one negative. Example: The square roots of 9 are 3 and –3, because $3^2 = 9$ and $(-3)^2 = 9$.	• no square root
• one cube root, which is positive. Example: The cube root of 8 is 2, because $2^3 = 8$.	• one cube root, which is negative. Example: The cube root of –8 is –2, because $(-2)^3 = -8$.

Exercise 2

1–6, 8

1 Write down the positive and negative square roots of:

a) 36 .... and ....

b) 16 .... and ....

c) 49 .... and ....

2 Write down the value of:

a) the negative square root of 64 ....

b) the cube root of –1000 ....

c) $\sqrt{81}$ ....

d) $\sqrt[3]{-64}$ ....

Think about

3 Why doesn't question 2 part d have two answers?

....

4 Tick to show if each statement is true or false.

	True	False
$\sqrt[3]{-8} = -2$	☐	☐
$\sqrt{-25} = -5$	☐	☐

5 Nikesh says that $6^3 - 6^2 = 6$.

Is this correct? Explain your answer.

....

6 Explain why it is not possible to find the square root of −16.

....

....

7 Use a calculator to find the values.

If an answer is not an integer, round it to two decimal places.

a) $\sqrt[3]{-343}$

b) $\sqrt[3]{-80}$

....

....

c) $\sqrt[3]{22.2} - \sqrt[3]{-18.6}$

d) $\sqrt[3]{13^2 - 14^2}$

....

....

8 Decide whether each statement is always true, sometimes true or never true.

a) The square root of an even number is even. ....

b) The cube of an even number is even. ....

c) The cube root of an odd number is odd. ....

1.3 The index laws

Summary of key points

The index laws

Use these laws to help you calculate with indices.

To multiply two powers of the same number, add the indices.	To divide two powers of the same number, subtract the indices.	To find a power of a power, multiply the indices.
Example: $3^4 \times 3^2 = 3^6$	Example: $8^9 \div 8^3 = 8^6$	Example: $(4^3)^5 = 4^{15}$

Exercise 3

1 Write as single powers:

a) $2^6 \times 2^3 = $ ....

b) $3^4 \times 3^3 = $ ....

c) $9^6 \div 9^2 = $ ....

d) $4^{12} \div 4^6 = $ ....

e) $(8^5)^2 = $ ....

f) $(5^6)^3 = $ ....

2 Simplify each expression. Leave your answer in index form.

a) $\dfrac{2^5 \times 2^7}{2^2}$ = ...

b) $\dfrac{7^9 \times 7^3}{7^5}$ = ...

c) $\dfrac{8^{11}}{8 \times 8^5}$ = ...

3 Write the missing power in each statement.

a) $6^{\square} \times 6^4 = 6^9$

b) $5^{13} \div 5^{\square} = 5^7$

c) $7 \times 7^{\square} \times 7^2 = 7^8$

d) $(4^{\square})^2 = 4^{12}$

e) $(4^{\square})^3 = 4^{21}$

4 Write the value of:

a) $3^{12} \div 3^9$

b) 8^0

c) $5^0 \times 5^1 \times 5^2$

....

5 Complete these number pyramids. The number in each cell is the product of the numbers in the two cells immediately below it.

a)
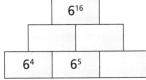

b)
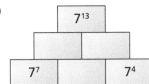

6 Moira says that $8^{10} \div 8^2$ simplifies to 8^5.

Explain the mistake that Moira has made.

....

....

7 The table shows some powers of 3.

Use the table to work out:

a) 243×81

....

b) $6561 \div 729$

....

$3^1 = 3$	
$3^2 = 9$	
$3^3 = 27$	
$3^4 = 81$	
$3^5 = 243$	
$3^6 = 729$	
$3^7 = 2187$	
$3^8 = 6561$	
$3^9 = 19\,683$	

2 2D and 3D shapes

You will practice how to:

- Identify and describe the hierarchy of quadrilaterals.
- Understand that the number of sides of a regular polygon is equal to the number of lines of symmetry and the order of rotation.
- Understand and use Euler's formula to connect the number of vertices, faces and edges of 3D shapes.
- Understand π as the ratio between a circumference and a diameter. Know and use the formula for the circumference of a circle.

2.1 Quadrilaterals

Summary of key points

The sum of the angles inside any quadrilateral is 360°.

The diagonals of a quadrilateral are the lines that connect one vertex of a quadrilateral to the opposite vertex.

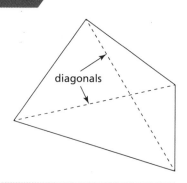
diagonals

> The diagonals of a quadrilateral bisect each other if one diagonal cuts the other exactly in half.

Here are some of the properties of special quadrilaterals.

Properties of a square

Four equal sides and angles.

Opposite sides are parallel.

Diagonals:

- are equal in length
- are perpendicular
- bisect each other.

Properties of a rectangle

Four equal angles.

Opposite sides are parallel and equal in length.

Diagonals:

- are equal in length
- bisect each other.

Properties of a rhombus

Four equal sides.

Opposite sides are parallel.

Opposite angles are equal.

Diagonals:

- are perpendicular
- bisect each other.

Properties of a parallelogram

Opposite sides are parallel and equal in length.

Opposite angles are equal.

Diagonals bisect each other.

Properties of a trapezium

One pair of sides are parallel.

Properties of an isosceles trapezium

One pair of sides are parallel.

The other pair of sides are equal in length.

Diagonals equal in length.

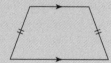

Properties of a kite

Two pairs of equal, adjacent sides.

One pair of opposite angles are equal.

Diagonals:

- are perpendicular and one diagonal bisects the other.

1 Write down the names of four different quadrilaterals that have opposite sides that are equal in length.

....

2 Tick the properties of each quadrilateral.

	Rectangle	Kite	Rhombus
One of more lines of symmetry	☐	☐	☐
All angles equal	☐	☐	☐
All sides equal	☐	☐	☐
Perpendicular diagonals	☐	☐	☐

3 Lucy is describing a quadrilateral:

The diagonals are perpendicular and equal in length.

She asks five of her friends to draw the quadrilateral she is thinking of.

a) Freddie draws this shape.

Explain why he is wrong.

....

....

b) Here are the shapes her other four friends draw.

Zak   Pete   Pria   Amy

Which friend has drawn the shape that Lucy described?.....................

4 Are these statements true or false?

	True	False
A rhombus is a special type of parallelogram.	☐	☐
A rhombus is a special type of square.	☐	☐
A square is a special type of rhombus.	☐	☐

5 The seven quadrilaterals in the box are sorted using the diagram.

Complete the diagram.

isosceles trapezium
rhombus   trapezium
parallelogram   rectangle
square   kite

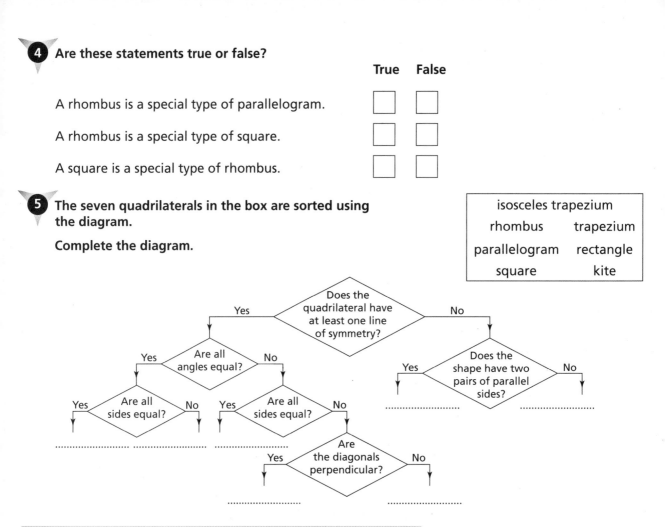

2.2 Polygon symmetry

Summary of key points

In a regular polygon:

• the number of sides is the same as the number of lines of symmetry
• and the number of sides is also the same as the order of rotational symmetry.

Exercise 2

1 Are these statements true or false?

	True	False
An equilateral triangle has 3 lines of symmetry.	☐	☐
A regular pentagon has rotational symmetry of order 5.	☐	☐
All hexagons have 6 lines of symmetry.	☐	☐
A regular octagon has rotational symmetry of order 8.	☐	☐

2 Choose words from the box to complete the sentences correctly. Words can be used more than once.

symmetry	pentagon	lines	heptagon	rotational	polygon

a) A regular .... has 7 .... of ....

b) A regular .... has the same number of .... of

.... as its order of ....

3 a) Sketch a regular polygon with 5 lines of symmetry and rotational symmetry of order 5.

b) Sketch a different shape with 5 lines of symmetry and rotational symmetry of order 5.

4 Explain what is wrong with each statement.

a) All hexagons have 6 lines of symmetry.

....

....

b) A regular nonagon has 8 lines of symmetry.

....

....

c) A regular decagon has rotational symmetry of order 12.

....

....

5 Ben draws a polygon with 7 sides.

Decide if each of the following statements could be true or cannot be true. Give a reason for each answer.

a) The polygon has no lines of symmetry.

Could be true ☐　　Cannot be true ☐

Reason: ....

....

b) The shape has rotational symmetry of order 7 and has 7 lines of symmetry.

Could be true ☐　　Cannot be true ☐

Reason: ....

....

2.3 3D shapes

Summary of key points

3D shapes have **vertices, edges** and **faces**.

A **vertex** is a point.

An **edge** is a line.

A **face** is a surface.

A **polyhedron** is a 3D shape with flat faces.

There are different types of **polyhedra**, such as **cuboids**, **pyramids** and **prisms**.

Euler's formula for all **polyhedra** is that $V + F - E = 2$

where　　V is the number of vertices

F is the number of faces

E is the number of edges.

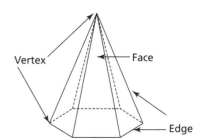

1 Explain how Euler's formula is illustrated in each of these 3D shapes.

a)	....
	....
	....
b)	....
	....
	....
c)	....
	....
	....

2 Complete this table for polyhedra.

	Number of vertices	Number of faces	Number of edges
Shape A	10		16
Shape B		10	12
Shape C	11	6	

3 Theo says that an icosahedron has 20 faces, 12 vertices and 28 edges.
He is incorrect about the number of edges.

a) How can you tell he is incorrect?

....

....

....

b) How many edges does a regular icosahedron have?

.... edges

Think about

4 Mae says she has made a polyhedron with 13 faces, 4 vertices and 11 edges. Amber tells her that she has the numbers of faces, vertices and edges mixed up.

The numbers are correct, but the names are mixed up.
What are the different possible correct combinations?

2.4 Circumference of a circle

Summary of key points

When you divide the circumference of any circle by the length of its diameter, you always get the same number.

This constant is called **pi (π)** and is approximately 3.14

The circumference, C, of a circle is found using the formula $C = πd$ or $C = 2πr$
where   d is the diameter and
            r is the radius of the circle.

Exercise 4

1 Find the circumference of a circle with each diameter. Round your answers to 1 decimal place.

a) 8 cm

.... cm

b) 11.8 cm

.... cm

c) 45 mm

.... mm

2 Draw a line to match each circle to the measurement of its circumference rounded to 1 decimal place.

6.5 cm

7.6 cm

5.2 cm

4.3 cm

13.5 cm   23.9 cm   27.0 cm   32.7 cm   40.8 cm   47.8 cm

3 A bicycle wheel has diameter 66 cm. Find how many turns the wheel makes when the bicycle travels 400 metres.

....

4 **a)** What is the length of the diameter of a circle with circumference 68 cm?
Give your answer to 1 decimal place.

.... cm

b) What is the length of the radius of a circle with circumference 45 cm?
Give your answer to 1 decimal place.

.... cm

5 A gardener uses a roller to flatten a bowling green of length 29 m.
The roller has a diameter of 1.3 m.

The gardener says the roller makes 8 full turns in travelling the length of the green. Is the gardener correct? Explain your answer.

....

....

....

....

Collecting data

You will practice how to:

- Select, trial and justify data collection and sampling methods to investigate predictions for a set of related statistical questions, considering what data to collect (categorical, discrete and continuous data).
- Understand the advantages and disadvantages of different sampling methods.

3.1 Data collection methods

Summary of key points

Primary data is information collected by the person who is doing the investigation.

Secondary data is information collected by someone else.

Exercise 1

1 Tick to show whether each person is using a primary or secondary data source.

	Primary data	Secondary data
Cara collects data about the prices of cars from a magazine.	☐	☐
Djillali asks all the people in his orchestra how long they spend practising their instrument each week.	☐	☐
Elsa records the temperature in her garden at 11:00 each day for a month.	☐	☐
Fahim collects data from a website about the number of visitors to some museums in 2017.	☐	☐

2 Thabisa wants to research the views of people in her village about plans for the village hall.

She decides to interview people face-to-face.

a) Are these statements about face-to-face interviews true or false?

	True	False
It is quicker than sending out a questionnaire.	☐	☐
It allows for more accurate answers as the interviewer can explain the questions.	☐	☐
It can give more reliable data as lots of people do not return questionnaires.	☐	☐

b) 2000 people live in the village. Put a ring around a suitable sample size for Thabisa's research.

   5   20   100   1000

3 Nick runs the supporters' club for a football team. He wants to know what the supporters think about the team's manager. He decides to send out a questionnaire to all supporters by email.

Are these statements about questionnaires true or false?

	True	False
It is more convenient than interviewing people face-to-face.	☐	☐
Data can be obtained quickly.	☐	☐
Most people will return the questionnaires.	☐	☐

4 Percy measures the speed of cars passing along a road.

Is Percy's information primary data or secondary data? Give a reason for your answer.

.... data because ....

....

....

5 A book gives the average life expectancy for people in different countries across all the world's continents.

a) Is this primary or secondary data? ....

b) Write down one question that Kosi could explore using the data in the book.

....

....

c) Why might Kosi get more accurate results if he collected the life expectancy data from the internet?

....

....

6 Tia wants to compare the amount of rainfall in 10 cities.

Which of these data collection methods do you think Tia should use? Give a reason for your choice.

Method 1	Method 2
Collect daily data about the amount of rainfall in each city from a weather website.	Set up her own experiments to record the amount of rainfall in each city.

....

....

....

Summary of key points

Data can be collected from everyone in the **population** you are interested in (but this could take a long time). Alternatively, data could be collected from a **sample**.

It is important that the sample used is **representative** of the population it is chosen from. A **random** selection method will help to make sure that the sample is representative (but this will not always be the most convenient way to collect the data).

Exercise 2

1 One thousand people watch a concert. André wants to know what people in the audience thought of the performance.

Tick whether these statements are true or false.

	True	False
It would take a long time to interview all the people in the audience.	☐	☐
It would be more convenient for André to collect data from a sample of the audience.	☐	☐
André could get reliable data by asking the first 10 people leaving the concert.	☐	☐

2 Natasha wants to find out what customers of a restaurant think of the food. She decides to interview people face-to-face.

Which method of choosing a sample should Natasha use? Give a reason for your choice.

A	B
Asking a sample of 20 customers using the restaurant one lunchtime.	Asking a sample of 10 customers using the restaurant on each of five separate occasions.

....

....

....

3 A hospital has 250 doctors and 750 nurses. Orla plans to interview a sample of 100 staff to find out how much they enjoy their work.

She decides to choose a sample of staff using one of these methods.

Method A	**Method B**	**Method C**
Sample 100 nurses eating in the canteen one afternoon.	Ask 75 doctors and 25 nurses chosen at random.	Ask 25 doctors and 75 nurses chosen at random.

Which method is best? Give reasons for your answer.

....

....

....

4 Simona wants to find out how much time children spend reading every day.

She considers these two methods of choosing a sample of children.

Method A	**Method B**
Ask a sample of 100 children using a library one morning.	Ask a sample of 20 children chosen at random from 5 different schools.

a) Write down one reason why Simona may want to use Method A.

....

....

b) Write down why Simona would get more reliable results using Method B.

....

....

....

5 Aarav wants to measure the heart rate of a sample of runners as they finish a marathon.

He considers measuring the heart rate of the first 50 runners to finish.

a) Give one reason why this may not be a convenient way to collect a sample of runners.

....

....

b) Suggest a better way for Aarav to collect his data.

....

....

....

4 Factors and rational numbers

You will practice how to:

- Understand factors, multiples, prime factors, highest common factors and lowest common multiples.
- Understand the hierarchy of natural numbers, integers and rational numbers.

4.1 Prime factors

Summary of key points

A whole number can be expressed as a product of its **prime factors**.

You can use a factor tree to find the **prime factorisation** of a number.

Example: This factor tree shows that the prime factorisation of $90 = 2 \times 3 \times 3 \times 5$.

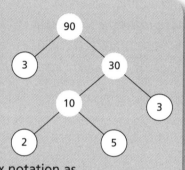

This can also be written using index notation as $90 = 2 \times 3^2 \times 5$.

You can find the **highest common factor** (HCF) and the **lowest common multiple** (LCM) of two numbers by writing the prime factors of each number.

For example, the prime factors of 18 are $2 \times 3 \times 3$.

The prime factors of 27 are $3 \times 3 \times 3$.

The HCF is the product of all the factors in common. The HCF of 18 and 27 is $3 \times 3 = 9$.	The LCM is the product of all the factors in common (once each), and all the other factors. The LCM of 18 and 27 is $2 \times 3 \times 3 \times 3 = 54$.

You can use a Venn diagram to show the HCF and LCM.

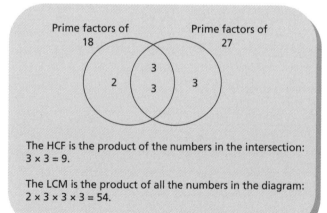

The HCF is the product of the numbers in the intersection:
$3 \times 3 = 9$.

The LCM is the product of all the numbers in the diagram:
$2 \times 3 \times 3 \times 3 = 54$.

Exercise 1

1 Draw a ring around each prime number.

41   42   43   44   45   46   47   48   49

2 Complete these factor trees.

a)

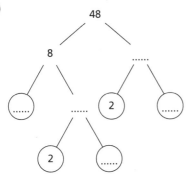

b)

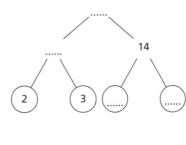

3 Write these prime factorisations using index notation.

a) $196 = 2 \times 2 \times 7 \times 7$ ....

b) $162 = 2 \times 3 \times 3 \times 3 \times 3$ ....

4 Write each number as a product of its prime factors, using index form.

a) 28

b) 88

....

....

c) 350

d) 520

....

....

5 **a)** Write 30, 42 and 63 as products of their prime factors.

30 = ... 42 = ... 63 = ...

b) What is the highest common factor of 30 and 42? ....

c) What is the highest common factor of 42 and 63? ....

d) What is the lowest common multiple of 30 and 42? ....

e) What is the lowest common multiple of 30 and 63? ....

6 **You are given that $132 = 2^2 \times 3 \times 11$ and $110 = 2 \times 5 \times 11$.**

a) Find the highest common factor of 132 and 110.

....

b) Find the lowest common multiple of 132 and 110.

....

c) Maxine says that 13 is a factor of 264.

Use the prime factorisation of 264 to explain why she is wrong.

....

....

7 **Olga and Carlos each buy some packets of pens. All the packets contain the same number of pens.**

Olga's packets contain a total of 96 pens.

Carlos' packets contain a total of 132 pens.

What is the largest possible number of pens that could be in a packet?

....

4.2 Rational numbers

Summary of key points

Rational numbers are numbers that can be written as fractions.

Examples: $-\frac{5}{2}$, $3 = \frac{3}{1}$, $0.71 = \frac{71}{100}$

Integers are whole numbers: ... −3, −2, −1, 0, 1, 2, 3, ...

The **natural numbers** are the positive integers: 1, 2, 3, 4, 5, ...

1 **a)** Write these labels in the correct places on the Venn diagram.

integers   natural numbers   rational numbers

....

....

....

b) Write these numbers in the correct places on the diagram.

8   −13 $3\frac{1}{2}$ 1   0.675 $-\frac{3}{11}$ π

2 **Write down a number that is:**

a) an integer but not a natural number   ....

b) a rational number but not an integer   ....

3 **Sajid says he can think of a natural number that is not rational.**

Is he correct? Explain your answer.

....

4 **Is each statement true or false?**

	True	False
All integers are rational numbers.	☐	☐
All integers are natural numbers.	☐	☐
All natural numbers are rational numbers.	☐	☐
All positive numbers are rational numbers.	☐	☐

5 **Write a rational number that lies between:**

a) −5 and −7   **b)** $\frac{1}{2}$ and $\frac{2}{3}$ **c)** −0.5 and $-\frac{1}{4}$

....

Think about

6 **Is it always possible to find a rational number that lies between two other rational numbers?**

5 Expressions

You will practice how to:

- Understand that letters have different meanings in expressions, formulae and equations.
- Understand that the laws of arithmetic and order of operations apply to algebraic terms and expressions (four operations, squares and cubes).
- Understand how to manipulate algebraic expressions including:
  - applying the distributive law with a single term (squares and cubes)
  - identifying the highest common factor to factorise.
- Understand that a situation can be represented either in words or as an algebraic expression, and move between the two representations (linear with integer or fractional coefficients).

5.1 Expressions, equations and formulae

Summary of key points

An **expression** connects numbers and **variables** with mathematical operations (such as $+$, $-$, $\times$ and $\div$).

For example:
$3x + 2y - 5$

An **equation** contains an equals sign, $=$. It is a statement that shows that the expressions on each side of the equals sign are equal.

For example:
$3 + 8 = 8 + 3$
$2x + 1 = x + 7$

In an equation, the letter represents a particular value (the **unknown**) that can be found by solving the equation.

A **formula** is a type of equation that shows the relationship between certain variable quantities.

For example:
area of rectangle = length × width
$$(A = ab)$$

$y = 3x + 1$ is also a **formula** that describes the relationship between the variables x and y. It is also sometimes known as a **linear function**.

y is called the **subject** of the formula.

1 Tick all the true statements.

A

In the equation
$2x + 3 = 11$, x is a
particular value that
can be found by
solving the equation.

B

In the formula $s = \dfrac{d}{t}$,
s, d and t all have
fixed values.

C

The value of the
expression
$y + 2\sqrt{x} - 3$ can be
found for any given
values of x and y.

D

In the formula
$A = a + y$,
y is called the subject
of the formula.

2 Draw a ring around all the equations.

$3 + 4 = 4 + 3$ $x \rightarrow 2x - 6$ $6x - 3 + y$

$nm + 4$ $2x - 4 = x$ $n = 28 - 2n$

3 Draw a ring around all the formulae.

$x \rightarrow 5x + 2$ $20 - 2n$ $4(x + 2) = 16$

Volume, $V = abc$ $4x - y + 7$ Cost = 5 × number of hours

4 Are these statements true or false?

	True	False
$3n + 2m$ is a formula.		
$t + 3 = 11$ is an expression.		
$P = 4x$ is a formula.		
$16 - n = 3n$ is an equation.		
$m + 2n$ is an expression.		

5 Arrange these formulae into two groups.

Explain why you grouped them the way you did.

$y = 3x + 1$ $A = 5m - 5$ $1 - 8x = A$

$A = 3x + 1$ $y = h + 1.9$ $10p + 1 = y$

....

....

....

5.2 Algebraic operations and substitution

Summary of key points

In $3x^2$, the power is calculated first.

In $3(a - 5)$ the bracket is calculated first.

In $\dfrac{2a + 7}{3}$ the numerator is calculated first.

Follow the rules of BIDMAS when substituting: work out brackets first, then powers, then division and multiplication followed by addition and subtraction.

Example: If $a = -5$ and $b = 4$ then

$$6b - 2a = (6 \times 4) - (2 \times -5)$$
$$= 24 - (-10)$$
$$= 34$$

$$2a^2 = 2 \times (-5)^2$$
$$= 2 \times 25$$
$$= 50$$

$$\frac{b^3 + 8}{b} = \frac{4^3 + 8}{4}$$
$$= \frac{64 + 8}{4}$$
$$= \frac{72}{4}$$
$$= 18$$

Exercise 2

1 Find the value of each expression when $a = 8$ and $b = 3$.

a) $5a + b$ ....   **b)** $a - 4b - 2$ ....   **c)** $\dfrac{a}{2} + 6b$ ....

d) b^2 ....   **e)** $4b^2$ ....   **f)** $a^2 - ab$ ....

2 Find the value of each expression when $p = 5$ and $q = -4$.

a) $36 - 9p$ ....   **b)** $pq + 35$ ....   **c)** $\dfrac{30}{p} - q$ ....

d) $p^3 - 20$ ....   **e)** $3q^2 + p$ ....   **f)** $2p^3$ ....

3 Draw a line to match each expression to its value when $m = 3$ and $n = 7$.

$9m - n$	9
$2m^2$	18
$2n - \dfrac{15}{m}$	20
$6n - 2m$	31
$n^2 - 6m$	36

4 Use the formula $y = 7x - 11$ to find the value of y for each value of x.

a) $x = 0$ $y = \ldots\ldots\ldots$

b) $x = -2$ $y = \ldots\ldots\ldots$

5 For each algebraic expression, underline the part of the calculation that should be performed *first*.

a) $5n + 4$ multiplication   addition

b) $t^3 + 4$ power   addition

c) $20 \div (a + 1)$ division   brackets

d) $30 - 4m$ subtraction   multiplication

6 Is the stated order of operations correct or incorrect?

	Correct	Incorrect
$6t - 3$ multiplication first, then subtraction	☐	☐
$(t + 1)^2$ brackets first, then square	☐	☐
$3 + t \div 2$ addition first, then division	☐	☐
$2t^2$ multiplication first, then square	☐	☐

7 Use the formula $p = 4a - 3b$ to find p when:

a) $a = 2$ and $b = -1$ $p = \ldots\ldots\ldots$

b) $a = -5$ and $b = 2$ $p = \ldots\ldots\ldots$

8 Use the formula $z = 5w + 2x$ to find:

a) z when $w = 3$ and $x = -2$

$z = \ldots\ldots\ldots$

b) x when $z = 27$ and $w = 3$

$x = \ldots\ldots\ldots$

c) w when $z = 43$ and $x = 4$

$w = \ldots\ldots\ldots$

9 Complete these statements for $t = -2$.

a) $4t + \ldots\ldots\ldots = 12$ b) $5t^2 - \ldots\ldots\ldots = 16$ c) $t^3 - \ldots\ldots\ldots = -13$

10 A prism has height h cm and a square base of side x cm.

The surface area, A cm^2, of the prism is given by $A = 2x^2 + 4xh$.

Find the value of h when $A = 192$ and $x = 6$.

$h = \ldots\ldots\ldots$

11 When n is an integer, are these statements always true, sometimes true or never true?

	Always true	Sometimes true	Never true
$8n + 24 = 0$	☐	☐	☐
$n^2 + 4 > 0$	☐	☐	☐
$n^2 - 4 = 0$	☐	☐	☐
$1 - n^2 > 2$	☐	☐	☐

Think about

12 $y = ax + b$ where a and b are positive whole numbers less than 20.

When $x = -2$, $y = 4$.

Find all the possible values of a and b.

Summary of key points

Brackets are **expanded** (or removed) by multiplying the terms inside the brackets by the expression on the outside. The reverse of expanding brackets is **factorising**.

Expanding brackets

$$y(2y + 3) \quad = \quad 2y^2 + 3y$$

Factorising

Examples:

$$14e + 21 = 7 \times 2e + 7 \times 3 \qquad 2a^2b - 8ab = 2ab \times a - 2ab \times 4$$
$$= 7(2e + 3) \qquad\qquad\qquad = 2ab(a - 4)$$

Exercise 3

1 **Match the equivalent expressions.**

| $6n + 10$ | $8n + 12$ | $6n + 12$ | $18n + 6$ |

| $6(3n + 1)$ | $4(2n + 3)$ | $6(n + 2)$ | $2(3n + 5)$ |

2 **Expand the brackets:**

a) $6(t + 4) = $ ....

b) $5(2m + n - 1) = $ ....

c) $4(5r + 4q) = $ ....

d) $n(n + 4) = $ ....

e) $m(m - 2) = $ ....

f) $2t(t - 3) = $ ....

g) $4n(2n + 5m + 1) = $ ....

h) $3ab(2a - 5b) = $ ....

3 **Draw a ring around all the correct statements.**

$4(z - 3) = 4z + 1$ $m(m + 4) = 2m + 4$ $n(n - 3) = n^2 - 3n$

$t(t + 3u) = t^2 + 3ut$ $6a(2a - 4) = 8a^2 + 2a$ $cd(2c - 3d + 1) = 2c^2d - 3cd^2 + cd$

4 Factorise:

    a) $a^2 + 5a$ ....

    b) $h^2 - 3h$ ....

    c) $9m - m^2$ ....

    d) $y^3 + 6y$ ....

5 Are these factorisations correct or incorrect?

	Correct	Incorrect
$6r^2 + 3r = 3r(2r + 1)$	☐	☐
$8mn - 12m = 8m(n - 4)$	☐	☐
$8w^2 + 4w = 4w(2w + 0)$	☐	☐
$12u^3 + 15u^2 = 3u^2(4u + 5)$	☐	☐

6 Complete these factorisations.

    a) $7v^2 + 14v =$ ....$(v +$ ....$)$ **b)** $10ef + 15fg = 5f($...........$+$ ....$)$

    c) $6p^2q - 15pq =$ ....$($...........$- 5)$ **d)** $12a^3b + 8ac = 4a($...........$+$ ....$)$

    e) $7n($...........$+ 4) = 21n^2 +$ ....   **f)** ....$(3n -$ ....$) = 12n^2 - 8n$

7 Factorise completely:

    a) $9u^2 - 18u$ ....   **b)** $4ab - a$ ....

    c) $8cd + 12c$ ....   **d)** $40f^2g - 32f$ ....

8 Factorise completely:

    a) $25u^3v - 30uv^2$ ....   **b)** $24m^2n - 8mn^2 - 32mn$ ....

9 Expand and simplify:

    a) $4(2m - 3) + 5(3 - 2m) =$ ....

    b) $3(2p - 3q + 1) + 2(p - 4 - q) =$ ....

Think about

10 Write down five different expressions that can be factorised to the form

    ....$(2a + 3b)$

5.4 Forming expressions

Summary of key points

Anya is a years old. Boris is 3 years older than Anya. Clara is twice as old as Boris.	→ Translate text to symbols →	Anya's age = a years Boris' age = a + 3 years Clara's age = 2(a + 3) years = 2a + 6 years

Anya is a years old.

Boris is 3 years older than Anya.

Clara is twice as old as Boris.

Translate text to symbols

Anya's age = a years

Boris' age = a + 3 years

Clara's age = 2(a + 3) years

$\quad\quad\quad = 2a + 6$ years

Find an expression for the sum of the ages of the three children.

a + a + 3 + 2a + 6

= 4a + 9 years

The sum of their ages is 4a + 9 years.

Exercise 4

1 The table shows some information about three books.

Book	Number of pages	Cost
Book A	a pages	$7
Book B	250 pages	$b
Book C	c pages	$6

Write down an expression for:

a) the total number of pages in the three books

.... pages

b) the difference between the cost of Book B and Book C

$ ....

c) the change a person should receive if they use a $50 note to buy one of each book.

$ ....

2 The table gives information about the masses of four pieces of fruit.

a) Complete the table by writing each mass in terms of a.

Fruit	Description	Expression for mass
Apple	The apple has mass a grams.	a grams
Lemon	The lemon has a mass that is 20 grams **less** than the mass of the apple.	.... grams
Grapefruit	The grapefruit's mass is 25 grams more than the mass of the apple.	.... grams
Melon	The melon's mass is three times the mass of the apple.	.... grams

b) Find the total mass of all four pieces of fruit. Give your answer in terms of a.

.... grams

3 The length of a rectangle is 4 cm more than its width. Given that the width of the rectangle is w cm, write an expression for the perimeter of the rectangle.

.... cm

4 Five children take a test.

Complete the table describing and comparing the children's marks.

Child	Description of marks	Expression for marks (in terms of x)
Eryl	Eryl's score was x.	x
Flyn	Flyn scored 5 marks less than Eryl.	....
Georgia	Georgia scored .... marks more than Flyn.	$x + 10$
Hafa	Hafa scored .... as many marks as Georgia.	$2x + 20$
Indira	Indira scored .... marks more than Hafa.	$2x + 30$

5 Raj buys 4 packets of biscuits for a meeting. Each packet contains b biscuits. After the meeting, there are only 7 biscuits remaining.

How many biscuits were eaten?

.... biscuits

6 Tina has *n* sweets. Max has 6 sweets. They combine their sweets and share them equally among 3 children.

How many sweets does each child get?

.... sweets

7 A mug has mass *w* grams. These mugs are packed into boxes of 12. The box has mass *b* grams.

Find an expression for the total mass of the box and mugs.

.... grams

8 A strip of wood has length *L* cm. Five pieces, each of length *x* cm, are cut from the strip.

Find an expression for the length of wood that remains.

.... cm

9 Petrol costs $*c* per litre and a bottle of engine oil costs $*e*.

Nina buys 30 litres of petrol and 2 bottles of engine oil.

Write an expression for the total cost.

$ ....

10 Sandra receives $*a*. She spends $*b* on a magazine. She spends one-third of the remaining money on a coat.

How much money does she spend on the coat?

$ ....

11 Jenson buys *t* books.

x of the books each cost $*a*. The remaining books each cost $*b*.

Write an expression for the total cost of the books.

$ ....

12 Using *n* to represent the original number in each case, write an expression for the final result.

a) I think of a number.

I multiply it by 4.

I add 6.

I then multiply by 2.

Finally I subtract 3.

....

b) I think of a number.
I add 1.

I then multiply by 4.

I add 6.

I multiply by 2.

Finally I add 8.

....

c) I think of a number.

I multiply it by 4 and subtract the result from 20.

....

d) I think of a number.

I subtract it from 10 and multiply the result by 4.

....

Think about

13 Make up your own 'think of a number' puzzle. Use a letter to represent your starting number and find an expression for your final result.

14 Write an expression in terms of *x* for the size of angle *ABC*.

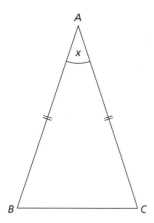

....

15 Frank makes *n* cakes.

25% of these are fruit cakes. Each fruit cake uses *a* grams of flour.
The rest of the cakes are chocolate cakes. Each chocolate cake uses *b* grams of flour.

Write an expression for the total amount of flour Frank uses.

.... grams

Angles

You will practice how to:

- Recognise and describe the properties of angles on parallel and intersecting lines, using geometric vocabulary such as alternate, corresponding and vertically opposite.
- Derive and use the fact that the exterior angle of a triangle is equal to the sum of the two interior opposite angles.
- Understand and use bearings as a measure of direction.

6.1 Special angles

Summary of key points

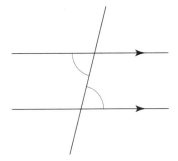

Alternate angles are equal.

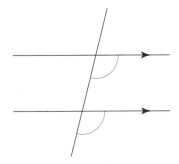

Corresponding angles are equal.

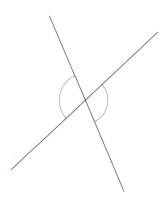

Vertically opposite angles are equal.

1 Write down whether the marked angles are alternate angles, corresponding angles or vertically opposite angles.

a)

....

b)

....

c)

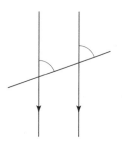

....

d)

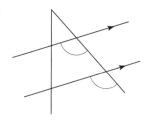

....

e)

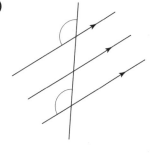

....

f)

....

2 Mark the angle described on each diagram.

a)

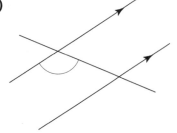

Angle A is corresponding to the marked angle.

b)

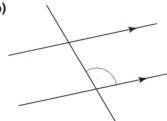

Angle B is alternate to the marked angle.

c)

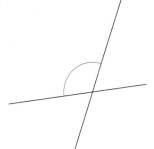

Angle C is vertically opposite to the marked angle.

d)

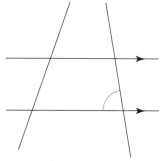

Angle D is alternate to the marked angle.

3 Mark all the angles that are equal to 60°.

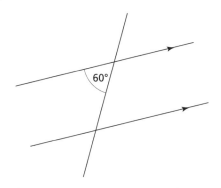

4 Are these statements true or false?

	True	False

a)

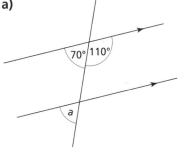

Angle a = 70° ☐ ☐

b)

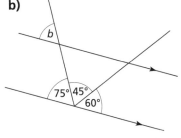

Angle b = 45° ☐ ☐

c)

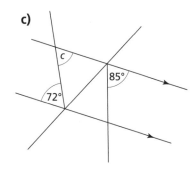

Angle c = 85° ☐ ☐

5 Write down the size of each lettered angle.

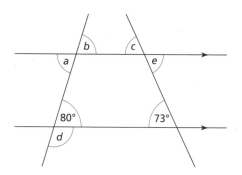

Angle a = ...°

Angle b = ...°

Angle c = ...°

Angle d = ...°

Angle e = ...°

6 Theo looked at this diagram and said,

'Angle *b* is 70° and angle *f* is 100°.'

Is he correct? Explain your answer.

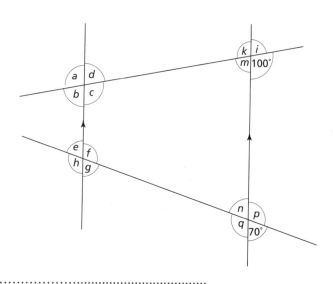

....

....

....

....

....

....

6.2 Angles of a triangle

Summary of key points

The angles inside a shape are called **interior** angles.

The angles outside a shape are called the **exterior** angles.

Angle w is one of the **exterior** angles of triangle ABC.

> Remember that the angles in a triangle add up to 180°.

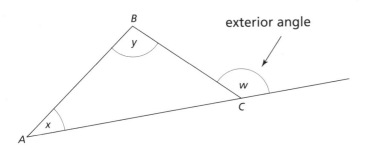

The exterior angle of a triangle is equal to the sum of the two interior opposite angles, that is w = x + y.

Exercise 2

1 Complete this proof about the exterior angle in a triangle.

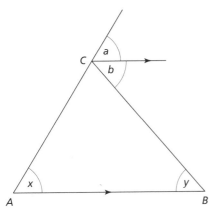

a = x because ....

b = y because ....

The exterior angle at C is a + b = x + y .

So the exterior angle at C is the sum of the interior angles at A and B.

2 Find the size of each lettered angle.

a)

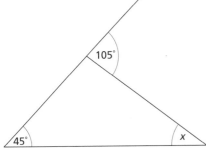

x = ...°

b)

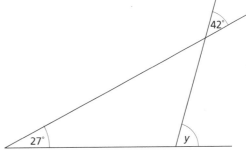

y = ...°

3 Find the value of *x* in each diagram. Give geometric reasons to support your answers.

a)

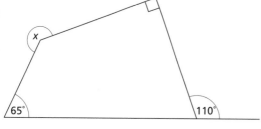

x = ...°

....

....

....

b)

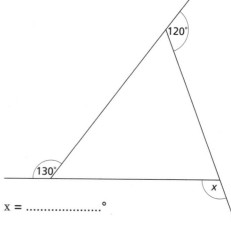

x = ...°

....

....

....

4 Find the size of angle *t*. Give geometric reasons for your answer.

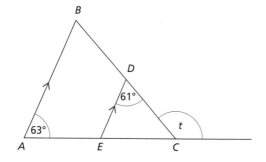

t = ...°

....

....

....

....

5 Find the values of *x* and *y*.

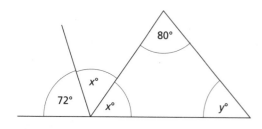

x = ...,   y = ...

6 The diagram shows a kite drawn inside an equilateral triangle.

Find the size of the angle marked *x*. Show how you worked out your answer.

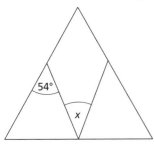

x = ...°

7 *AB* and *DF* are parallel lines. Find the size of the angles marked *x*, *y* and *z*.
Give geometrical reasons for each answer.

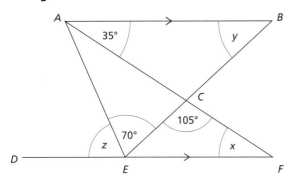

x = ...° because ....

y = ...° because ....

z = ...° because ....

8 This shape is made from two congruent squares and two isosceles triangles.
Amber said, 'The angle *a* is 138°.'

Is she correct? Explain your reasons.

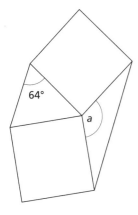

....

....

....

....

....

....

....

....

....

....

....

....

....

....

6.3 Bearings

Summary of key points

Bearings are angles measured clockwise from north. They measure direction and are written with 3 digits, for example 090° is due east.

When measuring or drawing a bearing, remember to begin by drawing a **north line** if it isn't already given.

Compass points

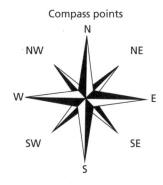

Exercise 3

1 **Choose the bearing that matches each direction.**

090°	180°	270°
100°	025°	045°
135°	050°	315°

a) west ....   b) south ....   c) north east ....

2 Measure the bearing of *Q* from *P*.

a)

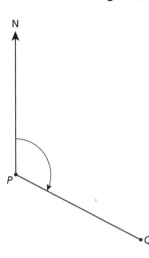

....

b)

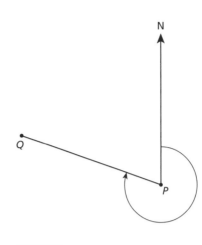

....

3 Draw these bearings.

a) The bearing from P is 065°.

b) The bearing from P is 257°.

4 The diagram shows the position of an airport (*A*) and a railway station (*R*).

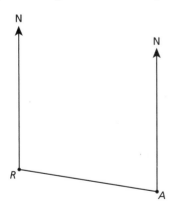

a) Measure the bearing of A from R.   ....°

b) Measure the bearing of R from A.   ....°

c) A plane leaves the airport and flies on a bearing of 050°.

Draw the path of the plane on the diagram.

5 The diagram shows the position of three towns, *L, M* and *N*.

L•

•M

•
N

Complete the statements.

a) The bearing of N from L is ....°.

b) The bearing of N from M is ....°.

c) The bearing of L from M is ....°.

6 The diagram shows the course for a boat race.

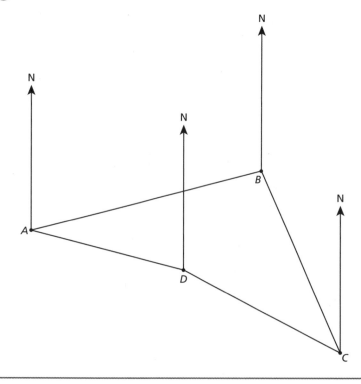

Complete the table to show the bearing for each stage of the course.

Stage	Bearing
From A to B	
From B to C	
From C to D	
From D to A	

7 The diagram shows the position of points *A* and *B*.

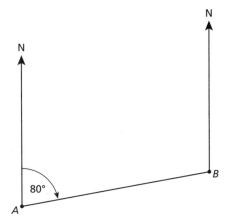

Eve says that the bearing of *B* from *A* is 80°. Explain why she is incorrect.

....

....

8 The bearing from *D* to *E* is 125°. What is the bearing from *E* to *D*?

.....................°

9 The diagram shows the position of a lighthouse (*L*) and a harbour (*H*).

The bearing of a boat (*B*) from *L* is 072°.

The bearing of the boat from *H* is 280°.

Mark the position of the boat on the diagram.

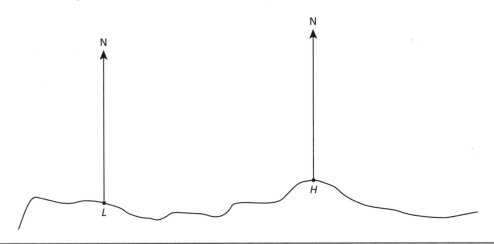

7 Place value, rounding and decimals

You will practice how to:

- Use knowledge of place value to multiply and divide integers and decimals by 0.1 and 0.01.
- Round numbers to a given number of significant figures.
- Estimate and multiply decimals by integers and decimals.
- Estimate and divide decimals by numbers with one decimal place.

7.1 Multiplying and dividing by 0.1 and 0.01

Summary of key points

Multiplying by 10 is the same as dividing by 0.1.
Multiplying by 100 is the same as dividing by 0.01.

Example:

$$3.34 \xrightarrow[\div\,0.01]{\times\,100} 334$$

Dividing by 10 is the same as multiplying by 0.1.
Dividing by 100 is the same as multiplying by 0.01.

Example:

$$65.7 \xrightarrow[\times\,0.1]{\div\,10} 6.57$$

Exercise 1

1 Work out:

a) 1050 × 0.01 ....

b) 4.5 × 0.01 ....

c) 24 682 × 0.1 ....

d) 0.028 × 0.01 ....

2 Work out:

a) 51 ÷ 0.1 ....

b) 216 ÷ 0.01 ....

c) 1.94 ÷ 0.1 ....

d) 2.74 ÷ 0.01 ....

e) 0.045 ÷ 0.1 ....

f) 0.0068 ÷ 0.01 ....

3 Join the parts to make four correct sentences.

Multiplying by 0.1 is the same as	dividing by $\frac{1}{100}$	or multiplying by 10.
Multiplying by 0.01 is the same as	multiplying by $\frac{1}{100}$	or dividing by 100.
Dividing by 0.1 is the same as	dividing by $\frac{1}{10}$	or dividing by 10.
Dividing by 0.01 is the same as	multiplying by $\frac{1}{10}$	or multiplying by 100.

4 Complete each statement by writing in one of these signs: < or > or =.

a) 28×0.01 28

b) $0.069 \div 0.01$ 6.9

c) $4.2 \div 0.01$ .... 4200×0.1

d) 0.048×0.01 .... $0.048 \div 0.1$

5 Complete the calculations by writing in one of these signs: × or ÷.

a) 35.2 .... 0.1 = 3.52

b) 65 .... 0.01 = 6500

c) 0.485 .... 100 = 48.5

d) 11 .... 0.1 = 110

e) 3.67 .... 0.01 = 0.0367

f) 0.034 .... 0.1 = 0.0034

6 Arrange these six cards to make two correct number statements.
Each card should be used only once.

| 0.017 | 0.1 | 17 | 0.01 | 170 | 1.7 |

$$\boxed{} \times \boxed{} = \boxed{}$$

$$\boxed{} \times \boxed{} = \boxed{}$$

7 Complete the missing number in each calculation.

a) 36 × .... = 3.6

b) 0.027 × .... = 27

c) 0.08 ÷ .... = 8

d) 2.56 × .... = 0.0256

8 A booklet has a mass of 0.01 kg. A pile of these booklets has a total mass of 0.8 kg.

How many booklets are in the pile?   ....

9 A snail travels 0.1 cm in one second. How far will it travel in 15 seconds?

.... **cm**

10 Kasim says that division always makes a number smaller.

Give an example to show that he is wrong.

....

7.2 Rounding to significant figures

The first **significant figure** of a number is the first non-zero digit.

Summary of key points

Example: Round 247 618 to 2 significant figures.	Example: Round 0.020 347 to 3 significant figures.
2 4 7 6 1 8 ↑ 2nd significant figure	0.0 2 0 3 4 7 ↑ 3rd significant figure
The next number (7) is large enough to round the number **up**.	The next number (4) is not large enough to round up.
247 618 → 250 000	0.020 347 → 0.0203
The 0s are needed at the end to preserve place value.	The 0s are needed at the start to preserve place value.

1 Draw a ring around:

a) the first significant figure in

b) the third significant figure in

c) the least significant figure in

$$0.02916 \qquad 0.003204 \qquad 123456$$

2 Round each number to 1 significant figure.

a) 4218 ....

b) 37 813 ....

c) 0.0247 ....

d) 0.004 718 ....

3 Round each number to the accuracy given in brackets.

a) 103 498 (3 significant figures) ....

b) 6 752 081 (2 significant figures) ....

c) 2.7842 (2 significant figures) ....

d) 0.305 287 (3 significant figures) ....

e) 49 702 (2 significant figures) ....

f) 0.007 024 (2 significant figures) ....

4 The Trans-Siberian Railway is 9289 km in length. Round this distance to 2 significant figures.

.... km

5 Use a calculator to do the calculations below. Round each answer to 3 significant figures.

a) $65 \div 8.7$

b) 5.31^3

c) $\dfrac{\sqrt{8.42+3.81}}{5.9}$

....

6 Ali says that 6.998 rounded to 2 significant figures is 7.

Explain why Ali is wrong.

....

What is the correct answer?   ....

7 Jack writes down an answer of 0.39 correct to 2 significant figures.

Draw a ring around the two numbers that could be his unrounded answer.

0.386   0.396   0.3845   0.394

8 Find an integer that is 1400 when rounded to 2 significant figures and 1450 when rounded to 3 significant figures.

....

Think about

9 Is a number rounded to 1 significant figure more or less accurate than the same number rounded to 2 significant figures?

7.3 Multiplying and dividing with integers and decimals

Summary of key points

Multiplying decimals

Example: 13.6 × 2.8

First complete the multiplication, ignoring the decimal points.

```
      1  3  6
×        2  8
  1  0  8  8
  2  7  2  0
  3  8  0  8
```

Then put the decimal point back in.

13.6 × 2.8 is equivalent to (136 × 28) ÷ 100, so the answer is 38.08.

Dividing decimals

Example: 5.85 ÷ 1.3

First find an equivalent calculation that does not involve dividing by a decimal.

5.85 ÷ 1.3 = 58.5 ÷ 13

Then carry out the division.

```
          4 .  5
13 ⟌ 5  8 . ⁶5
```

Remember to keep the decimal points aligned.

So the answer is 4.5.

Exercise 3

1 Find:

a) 0.3 × (−6)   b) −2 × 0.9   c) −8 ÷ 0.5   d) 3 ÷ 0.2

....

e) −40 ÷ 0.4   f) −7 ÷ 0.2   g) 0.22 × 4   h) −10 ÷ 0.4

....

2 Estimate and then calculate:

a) -56×0.4

b) $0.04 \times (-283)$

....

....

c) 0.26×0.38

d) 25.3×0.7

....

....

3 Complete this multiplication grid.

×	3.4	....
1.8	....	7.56
5.6	19.04	....

4 Grapes cost $4.40 per kilogram.

Estimate and then work out the cost of 0.8 kg of grapes.

$...........................

5 One orange costs $0.40. Mr Li spends $19.60 on oranges for his café.

Estimate and then calculate how many oranges he buys.

....

6 Estimate and then calculate:

a) $7.8 \div 0.3$

b) $-42 \div 0.7$

....

....

c) $4.48 \div 0.7$

d) $0.414 \div 0.9$

....

....

7 You are given that $-345 \div (-5) = 69$.

Are these statements true or false?

	True	False
$34.5 \div 5 = 6.9$	☐	☐
$-345 \div 0.5 = -6.9$	☐	☐
$34.5 \div 0.5 = 69$	☐	☐
$6.9 \times (-5) = 34.5$	☐	☐
$0.5 \times 0.69 = 0.0345$	☐	☐

 8 James says that $2.3^2 = 4.9$.

Describe the incorrect method James has used.

Show that his answer is incorrect.

....

....

....

Think about

9 Why does James' method give the wrong answer?

Presenting and interpreting data 1

You will practice how to:

- Record, organise and represent categorical, discrete and continuous data. Choose and explain which representation to use in a given situation:
  - o tally charts, frequency tables and two-way tables
  - o dual and compound bar charts
  - o frequency diagrams for continuous data
  - o stem-and-leaf diagrams.
- Interpret data, identifying patterns, trends and relationships, within and between data sets, to answer statistical questions. Discuss conclusions, considering the sources of variation, including sampling, and check predictions.

8.1 Frequency tables and diagrams

Summary of key points

Class intervals are used to group data in a frequency table. The **modal class** is the class interval that is most frequent.

Discrete data

Examples of grouped intervals for discrete data are:

1–4,   5–8,   9–12...

Notice that all intervals have the same width.

Continuous data

It is important to make sure there are no overlaps or gaps when choosing intervals for continuous data.

Such intervals are usually written using inequality signs, for example:

$10 \leq x < 20$ $20 \leq x < 30$ $30 \leq x < 40$...

Notice that all intervals have the same width.

The group $20 \leq x < 30$ includes any values that are at least 20 but less than 30.

A set of grouped data can be represented in a frequency diagram.

For grouped discrete data:

- The bars are equal width.
- There are gaps between the bars.
- Each interval is written below the corresponding bar.

For grouped continuous data:

- The bars are equal width.
- A continuous scale is used.
- There are no gaps between the bars.

1 Complete the intervals in each frequency table. Make sure the class intervals have equal widths.

a)

Interval	Frequency
5 ≤ x < 10	
10 ≤ x < 15	
15 ≤ x <	
...... ≤ x <	
..... ≤ x <	

b)

Interval	Frequency
0 ≤ x < 100	
100 ≤ x < 200	
.... ≤ x <	
.... ≤ x <	
.... ≤ x <	

2 Here are the masses, in grams, of some apples.

94   103   113   89   94   102   99   111   97   103

114   116   101   95   88   107   102   113   95   104

Mass, *m* (grams)	Tally	Frequency
80 ≤ m < 90		
90 ≤ m < 100		
100 ≤ m <		
.... ≤ m <		

a) Complete the first column. Make sure the class intervals have equal widths.

b) Complete the rest of the table.

c) How many of the apples weigh less than 100 grams? ....

3 The highest temperature (in °C) recorded each day in a town over a two-week period was:

18.6   15.2   12.8   18.3   20.4   16.3   16.6

17.1   19.5   13.4   15.7   16.7   21.5   19.6

a) Design a frequency table with five classes of equal width to record the data.

Temperature, T (°C)	Tally	Frequency
.... ≤ T <		
.... ≤ T <		
.... ≤ T <		
.... ≤ T <		
.... ≤ T <		

b) Complete the frequency table.

4 The table shows the number of teachers employed by 90 schools.

Number of teachers	Number of schools
0–9	9
10–19	16
20–29	37
30–39	18
40–49	10

Complete the frequency diagram for this set of data.

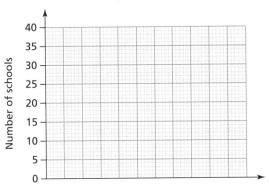

Number of teachers employed in schools

5 Luis measures the areas of 24 leaves. The table and the frequency diagram show his results.

Area, a (cm²)	Number of leaves
6 ≤ a < 8	
8 ≤ a < 10	
10 ≤ a < 12	6
12 ≤ a < 14	4
14 ≤ a < 16	1

Complete the frequency diagram and the table.

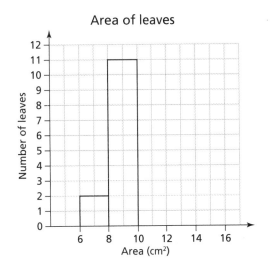

Area of leaves

6 The frequency diagram shows the amount of sleep (in hours) of a group of 80 people.

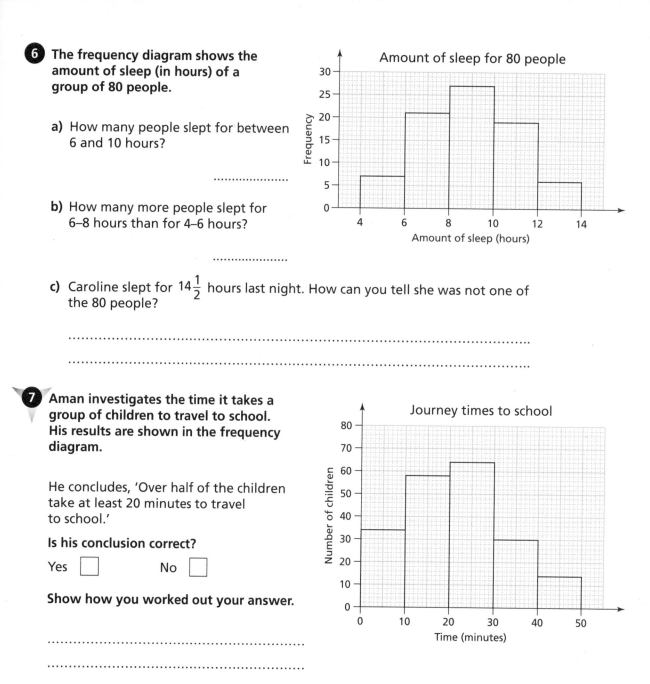

a) How many people slept for between 6 and 10 hours?

....

b) How many more people slept for 6–8 hours than for 4–6 hours?

....

c) Caroline slept for $14\frac{1}{2}$ hours last night. How can you tell she was not one of the 80 people?

....

....

7 Aman investigates the time it takes a group of children to travel to school. His results are shown in the frequency diagram.

He concludes, 'Over half of the children take at least 20 minutes to travel to school.'

Is his conclusion correct?

Yes ☐   No ☐

Show how you worked out your answer.

....

....

....

....

Summary of key points

Dual and **compound bar charts** can be drawn to compare two sets of data. For example, these diagrams compare the number of customers at a shop in the morning and in the afternoon.

Exercise 2

1. **The table shows the number of cars sold by three garages in May and June.**

	Garage A	Garage B	Garage C
May	46	18	23
June	27	32	28

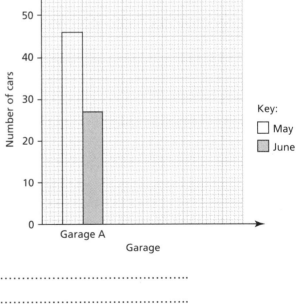

a) Complete the dual bar chart for this information.

b) Make two comparisons of the car sales between Garages A and B.

....

....

....

....

2 A cinema sells child and adult tickets.

The compound bar chart shows the percentage of each type of ticket sold by the cinema on three days.

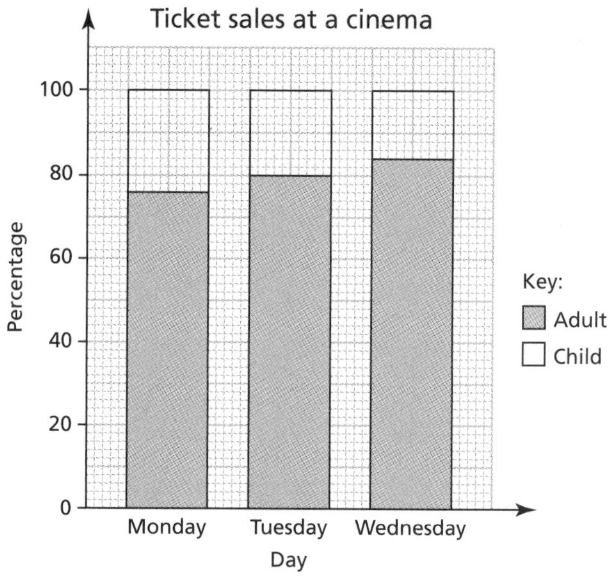

Ticket sales at a cinema

Percentage (y-axis): 0, 20, 40, 60, 80, 100

Day (x-axis): Monday, Tuesday, Wednesday

Key:
- ▉ Adult
- ☐ Child

a) Tick to show if each of these conclusions is true or false, or whether it is not possible to tell.

	True	False	Cannot tell
More adult tickets than child tickets were sold on Monday.	☐	☐	☐
More adult tickets were sold on Tuesday than were sold on Monday.	☐	☐	☐
84% of tickets sold on Wednesday were child tickets.	☐	☐	☐
On Tuesday, the ratio of child tickets to adult tickets sold was 1 : 4	☐	☐	☐

b) Comment on how the percentage of child tickets sold changed over the three days.

....

....

....

c) Give a reason why a compound bar chart is an appropriate form of diagram to show the data.

....

....

....

3 Kamal collects information about the favourite subjects of students in Year 7 and Year 8 in his school. He uses a random sample of 100 Year 7 students and 100 Year 8 students.

The dual bar chart shows some of his results.

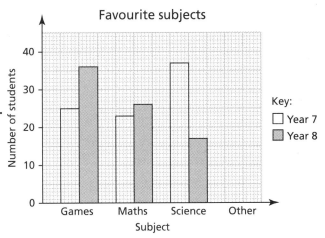

Favourite subjects

Key:
☐ Year 7
▦ Year 8

a) Complete the dual bar chart by drawing the bars for 'Other'.

b) Find how many more Year 7 students in the sample said 'Science' than Year 8 students.

....

c) Make two comparisons of the favourite subjects of students in Year 7 and Year 8.

....

....

d) The total number of students in Year 7 is 240.

Show that an estimate for the total number of students in the whole of Year 7 whose favourite subject is Games is 60.

4 In a survey, some men, women and children were asked how often they listened to music.

The results are shown in the compound bar chart.

How often people listen to music

Key:
☐ Children
▦ Men
▮ Women

a) Find the total number of women who took part in the survey.

....

b) Circle the modal response for children.

every day   most days   a few times each week   never

c) Find the number of men who said they never listen to music.

....

d) Compare the number of men and women who said they listen to music every day.

....

....

8.3 Stem-and-leaf diagrams

Summary of key points

Stem-and-leaf diagrams can be used to show ungrouped data values.

Each number is split into two parts – one part forms the stem and the other gives a leaf.

Exercise 3

1 Draw a stem-and-leaf diagram to show each set of data. Remember to complete the keys.

a) 11   14   21   25   26   27   32

    34   35   36   39   40   42

1			
2			
3			
4			

Key	|	represents

b) 141   137   125   145   130   152

    134   144   129   153   146   137

12			
13			
14			
15			

Key	|	represents

c) 5.2   5.6   7.1   4.6   6.1   7.3   4.9

    5.5   6.5   6.0   4.5   6.7   6.3   5.9

4			
5			
6			
7			

Key	|	represents

2 Here are the marks that 22 students scored in a test.

34   25   8   17   21   36   21   42   22   15   34

32   30   15   24   36   22   29   18   32   19   39

a) Draw a stem-and-leaf diagram to show the information.

b) Find the median mark.

....

Key:

c) Find the range of the marks.

....

3 Rick records the number of pizzas he sells each day.

The stem-and-leaf diagram shows his results.

```
0 | 4 9
1 | 0 1 3 5 7 8
2 | 1 2 5 6 6 7 9
3 | 0 2 3 3 8
4 | 1 5 5 7
5 | 0 0 0
```

Key 2 | 1 represents 21 pizzas

a) For how many days did he record this information? ....

b) On how many of the days does he sell 33 pizzas? ....

c) What is the smallest number of pizzas he sells in a day? ....

d) On how many days does Rick sell more than 30 pizzas? ....

e) Write down the modal number of pizzas he sells in a day. ....

f) Find the median number of pizzas he sells in a day. ....

4 Aisla draws a stem-and-leaf diagram to show the temperature at midday on 16 days.

Write down what is wrong with how Aisla has drawn her diagram.

```
15 |   2 4
16 |   0 6     3   9
17 | 5       8   2   1   0
18 | 5     7   1   4     8   7
19 |   6   1     4
```

....

....

....

Think about

5 A set of data has

- **15 values**
- **a median of 29**
- **a range of 42.**

Draw a possible stem-and-leaf diagram that matches these three conditions.

9

Functions and formulae

You will practice how to:

- Understand that a function is a relationship where each input has a single output. Generate outputs from a given function and identify inputs from a given output by considering inverse operations (including fractions).
- Understand that a situation can be represented either in words or as a formula (mixed operations), and manipulate using knowledge of inverse operations to change the subject of a formula.

9.1 Functions

Summary of key points

Here is an input–output table for the function $x \rightarrow \dfrac{x-3}{2}$.

Input, x		Output, y
1	→	−1
2	→	−0.5
3	→	0
4	→	0.5
5	→	1

This function can also be shown as a **mapping diagram**.

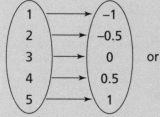 or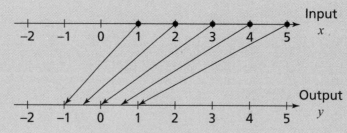

Exercise 1

1 a) Complete the function machine for $x \rightarrow 2x + 3$.

b) Complete the input–output table for the same function.

Input, x		Output, y
1	→	
2	→	
3	→	
4	→	

c) Show the mappings between these input and output values on the diagram.

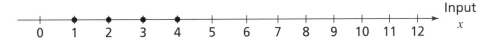

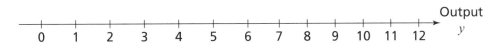

2 Complete the mapping diagram for this function machine.

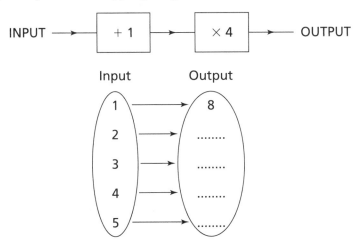

3 Complete the table using this function machine.

Input	Output
−3	
4.2	
	11
	14.8

4 Complete the table using this function machine.

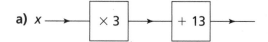

INPUT ⟶ [× 4] ⟶ [+ 7] ⟶ OUTPUT

Input	Output
$-1\frac{1}{2}$	
$\frac{1}{8}$	
	13
	−9

5 Find an expression in terms of x for the output of each function machine.

a) x ⟶ [× 3] ⟶ [+ 13] ⟶ $x \rightarrow$ ....

b) x ⟶ [÷ 2] ⟶ [− 6] ⟶ $x \rightarrow$ ....

c) x ⟶ [+ 3] ⟶ [× 8] ⟶ $x \rightarrow$ ....

d) x ⟶ [− 2] ⟶ [÷ 4] ⟶ $x \rightarrow$ ....

e) x ⟶ [× 2] ⟶ [+ 1] ⟶ [÷ 3] ⟶ $x \rightarrow$ ....

6 Complete the following so that the formulae and function machines match.

a) x ⟶ [.........] ⟶ [.........] ⟶ y $y = \frac{x}{4} - 11$

b) x ⟶ [.........] ⟶ [.........] ⟶ y $y = 5(x - 4)$

c) x ⟶ [÷ 6] ⟶ [.........] ⟶ y $y = \frac{x}{.......} + 3$

d) x ⟶ [.........] ⟶ [.........] ⟶ y $y = \frac{x - 14}{3}$

e) x ⟶ [.........] ⟶ [.........] ⟶ [× 2] ⟶ y $y = ...\left(\frac{x}{3} + 1\right)$

f) x ⟶ [× 4] ⟶ [+ 3] ⟶ [÷ .....] ⟶ y $y = \frac{.............}{6}$

7 Write or complete the algebraic expression for each mapping diagram.

a)

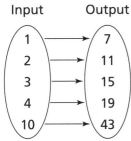

$x \rightarrow 4x +$ ....

b)

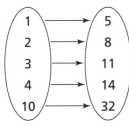

$x \rightarrow$ .... $x + 2$

c)

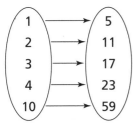

$x \rightarrow$ .... $x -$ ....

d)

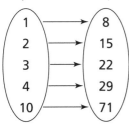

$x \rightarrow$ ....

e)

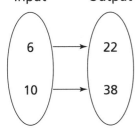

$x \rightarrow$ ....

8 Draw lines to match each function machine to the correct function.

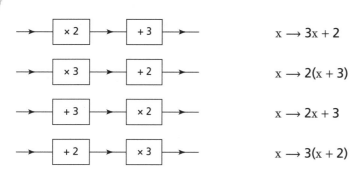

$x \longrightarrow 3x + 2$

$x \longrightarrow 2(x + 3)$

$x \longrightarrow 2x + 3$

$x \longrightarrow 3(x + 2)$

9 Here are two function machines.

a) Find the output from both function machines when the input is 7.

.... and ....

b) Find the output from both function machines when the input is $\frac{1}{2}$.

.... and ....

c) Use algebra to show that the output from both function machines will be the same for all input values.

....

....

....

Think about

10 Find five different functions that match this mapping diagram.
Write each function algebraically.

Input   Output

0.5 ⟶ 2

9.2 Constructing and using formulae

Summary of key points

An internet book store sells books for $6 each.

The store charges $4 for delivering each order.

A **word formula** to find the cost (in dollars) of buying any number of books and getting them delivered is:

cost (C) = 6 × number of books (n) + 4

This can be written as:

C = 6n + 4

The cost for ordering 7 books is:

C = 6 × 7 + 4

= 46

So the cost for ordering 7 books is $46.

Exercise 2

1 Find a formula for the perimeter, *P* cm, of this shape.

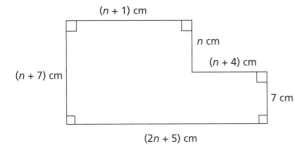

P = ...

2 A photographer charges $50 for each photograph a customer orders, plus a fixed charge of $100.

Find a formula for the total cost, $C, when a customer orders *n* photographs.

C = ...

3 The cross section of a prism is a polygon with *n* sides.

a) Write a formula for the number of faces, F, in terms of n.

F = ...

b) Use your formula to work out the number of faces when the cross section of the prism has 7 sides.

....

4 Drivers must pay $3 to cross a bridge in a car and $5 to cross a bridge in a van.

a) Work out the total money paid by 11 car drivers and 2 van drivers to cross the bridge in their vehicles.

$...........

b) Write a formula to find the total cost, $C, for c car drivers and v van drivers to cross the bridge in their vehicles.

C = ...

5 Biscuits are sold in two pack sizes, small and large.

contains b biscuits   contains B biscuits

Write a formula for the total number of biscuits, *N*, in *p* small packets and *P* large packets.

N = ...

6 Lottie finds these instructions on the internet for working out the surface area of a prism.

Step 1: Multiply the perimeter of the base by the height of the prism.

Step 2: Add on twice the area of the base.

a) Write a formula for the surface area of a prism, S, with base perimeter p, height h and base area A.

S = ...

b) Using your answer to part **a**, work out the surface area of a prism with p = 20, h = 6 and A = 24.

....

7 Photographs can be printed in two sizes, small and large.
The charges, $C, to print and deliver x small photographs and y large
photographs at two stores are given below.

Store 1	Store 2
$C = 0.3x + 0.7y$	$C = 0.2x + 0.5y + 5$

Kendra wants to have 20 small photographs and 10 large photographs printed
and delivered.
Which store has the lower cost for her order? Show how you worked out your
answer.

....

9.3 Changing the subject

Summary of key points

Given the length (l), width (w) and height (h) of a cuboid, the formula for the volume (V) is

$$V = lwh$$

The **subject** of this formula is V as the formula can be used directly to calculate
V from given values of l, w and h.

The subject of a formula can be changed.

Example:

Make x the subject of $y = 4x + 9$

$$y = 4x + 9$$

$$-9 \downarrow \qquad \downarrow -9$$

$$y - 9 = 4x$$

$$\div 4 \downarrow \qquad \downarrow \div 4$$

$$\frac{y - 9}{4} = x$$

So, the formula is: $x = \dfrac{y - 9}{4}$

1 Make r the subject of each formula.

a) $y = r - 5$

b) $y = 4r$

r = ...

r = ...

c) $y = r + d$

d) $y = \dfrac{r}{7}$

r = ...

r = ...

2 The cost of a taxi journey based on a \$4 fixed charge and \$3 per kilometre is given by the formula

$C = 3k + 4$

Make *k* the subject.

k = ...

3 The formula $L = 7P - 4$ is arranged to make *P* the subject.

Draw a ring around the correct formula.

$P = \dfrac{L}{3}$ $P = \dfrac{L-4}{7}$ $P = \dfrac{L+4}{7}$ $P = \dfrac{L-7}{4}$

4 Match the pairs of equivalent formulae.

$y = 2x + 5$	$y = 5x - 2$	$y = 2(x + 5)$	$y = \dfrac{x}{2} - 5$
$x = \dfrac{y + 2}{5}$	$x = 2(y + 5)$	$x = \dfrac{y - 5}{2}$	$x = \dfrac{y}{2} - 5$

5 Make x the subject of each formula.

a) $y = 3x - 1$

b) $y = ax + 8$

$x = $ ....

$x = $ ....

c) $y = 4(x - 3)$

d) $y = \dfrac{x + 9}{3}$

$x = $ ....

$x = $ ....

6 Pablo is trying to rewrite the formula $m = 4(n + 7)$ so n is the subject. Here is his working:

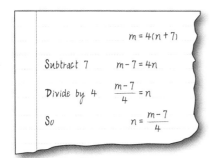

$$m = 4(n + 7)$$

Subtract 7 $\quad m - 7 = 4n$

Divide by 4 $\quad \dfrac{m - 7}{4} = n$

So $\quad n = \dfrac{m - 7}{4}$

What mistake has Pablo made?

....

....

10 Fractions

You will practice how to:

- Estimate and subtract mixed numbers and write the answer as a mixed number in its simplest form.
- Estimate and multiply an integer by a mixed number, and divide an integer by a proper fraction.

10.1 Subtracting mixed numbers

Summary of key points

If the first fraction is bigger than the second fraction	If the first fraction is smaller than the second fraction
Example: $2\frac{3}{4} - 1\frac{1}{5}$	Example: $3\frac{1}{3} - 1\frac{5}{6}$
Subtract the integers and the fractions separately:	Write one of the 1s of the first number as part of the fraction:
$2 - 1 = 1 \qquad \frac{3}{4} - \frac{1}{5} = \frac{15}{20} - \frac{4}{20} = \frac{11}{20}$	$2\frac{4}{3} - 1\frac{5}{6}$
Add the results:	Then subtract the integers and fractions separately:
$1\frac{11}{20}$	$2 - 1 = 1 \qquad \frac{4}{3} - \frac{5}{6} = \frac{8}{6} - \frac{5}{6} = \frac{3}{6} = \frac{1}{2}$
	Add the results:
	$1\frac{1}{2}$

An alternative method is to convert both numbers to improper fractions before subtracting.

Exercise 1

1. **Estimate and then calculate each answer.**

   If an answer is greater than 1, write it as a mixed number. Write all fractions in their simplest form.

   a) $1\frac{4}{5} - \frac{3}{10} =$ b) $3\frac{1}{2} - 1\frac{1}{3} =$ c) $3\frac{1}{3} - \frac{2}{5} =$

   ....

d) $2\frac{1}{4} - 1\frac{5}{6} =$ **e)** $3\frac{2}{5} - 1\frac{7}{10} =$ **f)** $4\frac{1}{2} - 2\frac{5}{7} =$

....

2 Match the cards with the same answer.

A $3\frac{2}{3} + 1\frac{1}{4}$

B $2\frac{1}{5} + 1\frac{1}{3}$

C $2\frac{1}{12} + \frac{5}{6}$

D $5\frac{1}{5} - 1\frac{2}{3}$

E $6\frac{1}{4} - 1\frac{1}{3}$

F $6\frac{2}{3} - 3\frac{3}{4}$

3 Nadia has $2\frac{1}{5}$ kg of flour. She uses $1\frac{1}{4}$ kg of it to make some cakes.

Estimate and then calculate how much flour she has left. Give your answer as a fraction in its simplest form.

.... kg

4 The difference between two fractions is $1\frac{4}{5}$. One of the fractions is $3\frac{1}{4}$.

Find the two possible values of the second fraction.

....

5 Use each of the digits 2, 3, 4, 5, 6 and 8 only once to complete this calculation.

$$\square\frac{\square}{\square} - \square\frac{\square}{\square} = 3\frac{23}{40}$$

Summary of key points

Multiplying an integer by a mixed number

Example: $3 \times 1\frac{4}{5}$

Partition the mixed number:

$$1\frac{4}{5} = 1 + \frac{4}{5}$$

Then use the **distributive** law to multiply the integer by each part:

$$(3 \times 1) + (3 \times \frac{4}{5}) = 3 + \frac{12}{5}$$

$$= \frac{27}{5} \text{ or } 5\frac{2}{5}$$

An alternative method is to convert the mixed number to an improper fraction and then multiply.

Dividing an integer by a fraction

Invert the fraction to make its **reciprocal** and then multiply:

Example: $4 \div \frac{2}{3} = 4 \times \frac{3}{2} = \frac{12}{2} = 6$

This gives the same answer as the original division.

Exercise 2

1 Estimate and then calculate each answer. Write your answers as mixed numbers with the fractions in their simplest form.

a) $3 \times 2\frac{1}{5}$ b) $4\frac{2}{3} \times 2$ c) $4 \times 1\frac{7}{8}$

....

2 Estimate and then calculate each answer. Write your answers as improper fractions in their simplest form.

a) $2 \times 1\frac{3}{4}$ b) $1\frac{2}{7} \times 4$ c) $3 \times 4\frac{1}{2}$

....

3 Find:

a) $5 \div \frac{1}{2}$ ....   b) $3 \div \frac{1}{4}$ ....

c) $6 \div \frac{1}{5}$ ....   d) $2 \div \frac{1}{8}$ ....

4 Work out each division and use a multiplication to check your answer. The first one has been done for you.

Calculation   Check

a) $4 \div \frac{1}{3} = 4 \times \frac{3}{1} = 12$ $12 \times \frac{1}{3} = 4$ ✓

b) $7 \div \frac{1}{4} =$ ....

c) $9 \div \frac{3}{4} =$ ....

d) $12 \div \frac{4}{5} =$ ....

e) $6 \div \frac{2}{3} =$ ....

Think about

5 Why are all of the answers to question 4 integers?

6 Write three examples of dividing an integer by a fraction where the answer is not an integer.

....

....

....

7 Write the missing integer in each box.

a) $\boxed{} \times 1\frac{1}{3} = 4$

b) $1\frac{\boxed{}}{5} \times 2 = 3\frac{3}{5}$

c) $\boxed{} \div \frac{1}{3} = 3$

d) $\boxed{} \div \frac{1}{3} = 6$

e) $3 \div \frac{1}{\boxed{}} = 21$

f) $4 \div \frac{\boxed{}}{5} = 10$

Length, area and volume

You will practice how to:

- Know that distances can be measured in miles or kilometres, and that a kilometre is approximately $\frac{5}{8}$ of a mile or a mile is 1.6 kilometres.
- Use knowledge of rectangles, squares and triangles to derive the formulae for the area of parallelograms and trapezia. Use the formulae to calculate the area of parallelograms and trapezia.
- Use knowledge of area and volume to derive the formula for the volume of a triangular prism. Use the formula to calculate the volume of triangular prisms.
- Use knowledge of area, and properties of cubes, cuboids, triangular prisms and pyramids to calculate their surface area.

11.1 Converting between miles and kilometres

Summary of key points

A **mile** is an **imperial unit** of length used to measure large distances.

These function machines can be used to give approximate conversions between miles and kilometres.

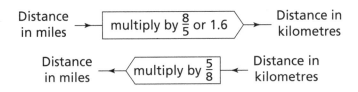

You can often use this relationship to convert between miles and kilometres:

5 miles is approximately equal to 8 kilometres.

Remember you can also convert **kilometres** to **miles** by multiplying by **0.625**.

Exercise 1

1 **Complete these statements:**

  a) 45 miles is approximately .... km

  b) 60 miles is approximately .... km

2 **Complete these statements:**

  a) 80 km is approximately .... miles

  b) 104 km is approximately .... miles

3 The distances on this road sign should be given in miles and kilometres.
Find the approximations for the missing values and complete the sign.
Give values to the nearest mile.

Rotterdam	152 km	.... miles
Nijmegen	32 km	.... miles

4 Draw a ring around the longer distance in each pair.

a) 56 miles   85 km   b) 504 miles   815 km

5 Complete these distance conversions.

a) 515 miles ≈ .... km   b) .... miles ≈ 424 km

c) 1.25 miles ≈ .... km   d) .... miles ≈ 7.2 km

6 On Monday, Ben drives 255 miles.
On Tuesday he drives 433 km.
Approximately how much further does he drive on Tuesday than on Monday?
Give your answer in kilometres.

.... km

7 Amber has enough petrol in her van to travel 43 miles.
The nearest petrol station is 65 km away.
Does Amber have enough petrol to drive to the petrol station?
Show how you worked out your answer.

....

....

....

8 When Matt drives on a motorway, his car uses 1 litre of petrol every 10 miles.
When he drives on any other type of road, his car uses 1 litre of petrol every 12 miles.

On a journey, Matt drives:
        240 km on motorways and
        96 km on other roads.
Work out how much petrol his car uses altogether.

.... litres

Think about

9 Find out about nautical miles. Where would you use nautical miles and how do they differ from an ordinary mile?

11.2 Area of 2D shapes

Summary of key points

Areas of shapes can be found using formulae.

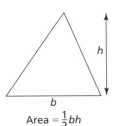

Area $= \frac{1}{2}bh$

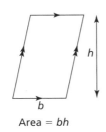

Area $= bh$

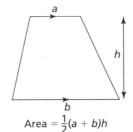

Area $= \frac{1}{2}(a + b)h$

Exercise 2

The shapes in this exercise are not drawn to scale.

1 Calculate the area of each shape.

a)

10 cm 8 cm

6 cm

.... cm²

b)

9 mm

5 mm

11 mm

.... mm²

2 Calculate the area of each shape, then draw a ring around the odd one out in each set.

a)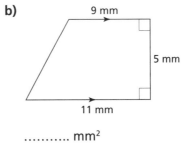

16 cm

18 cm

12 cm

12 cm

9 cm

24 cm

15 cm

....

b)

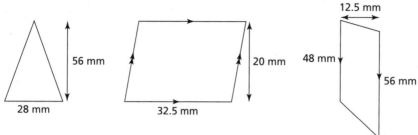

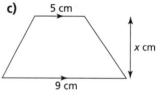

....

3 Find the value of *x* in each shape.

a)

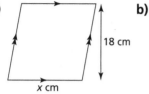

Area = 414 cm²

x = ...

b)

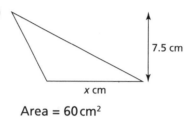

Area = 60 cm²

x = ...

c)

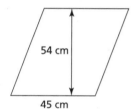

Area = 56 cm²

x = ...

4 The area of the parallelogram is the same as the area of the trapezium. Calculate the value of *h*.

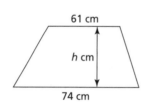

h = ...

5 Yan says that the area of the trapezium is more than 50% of the area of the square.

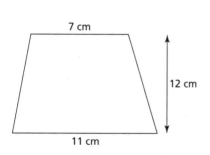

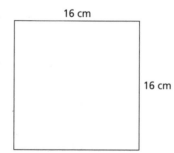

Is Yan correct?

Yes ☐   No ☐

Show how you worked out your answer.

....

....

....

11.3 Volume of triangular prisms

Summary of key points

Volume of a triangular prism = area of triangular cross section × length

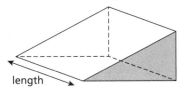

length

Exercise 3

1 Work out the volume of each triangular prism.

a)

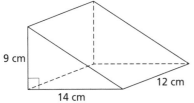

9 cm

14 cm

12 cm

.... cm³

b)

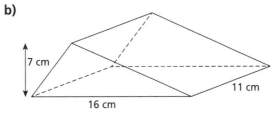

7 cm

16 cm

11 cm

.... cm³

2 Find the difference between the volumes of these two prisms.

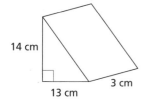

14 cm

13 cm

3 cm

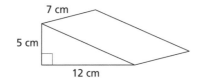

7 cm

5 cm

12 cm

.... cm³

3 The two prisms have the same volume. Find the height of prism B.

A

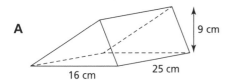

16 cm

25 cm

9 cm

B

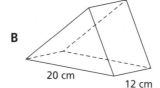

20 cm

12 cm

.... cm

4 Georgia looks at the prism and cuboid below. She says that the volume of the cuboid is 8 cm³ greater than the volume of the prism.

Is she correct? Show how you know.

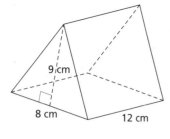

9 cm

8 cm

12 cm

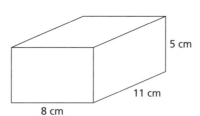

5 cm

11 cm

8 cm

....

....

....

....

Summary of key points

We can use nets to help us find the surface area of 3D shapes.

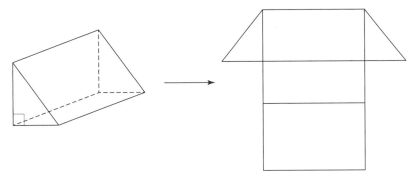

Exercise 4

1 Work out the surface area of each cuboid and then write them in the correct column in the table.

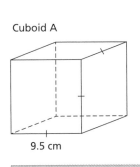

Cuboid A

9.5 cm

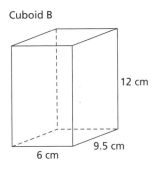

Cuboid B

12 cm

6 cm

9.5 cm

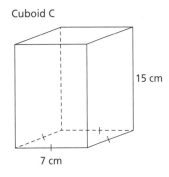

Cuboid C

15 cm

7 cm

Surface area less than 500 cm²	Surface area greater than 500 cm²

2 The diagram shows a net of a triangular prism drawn on a 1 cm square grid.

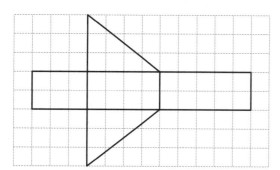

The shapes in this exercise are not drawn to scale.

Work out the surface area of the prism.   .... cm²

3 Each part shows the net of a solid shape. Work out the surface area of the solid.

a) Triangular prism

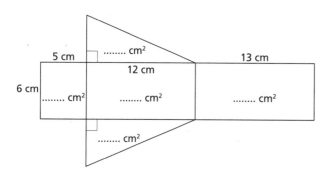

5 cm   .... cm²
        12 cm   13 cm
6 cm
.... cm²   .... cm²   .... cm²

.... cm²

.... cm²

b) Triangular prism

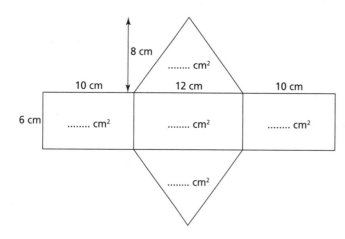

8 cm
.... cm²
10 cm   12 cm   10 cm
6 cm
.... cm²   .... cm²   .... cm²

.... cm²

.... cm²

4 Work out the surface area of the prism that is formed from this net.

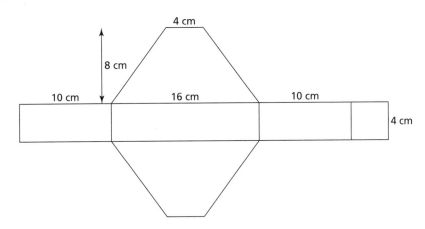

.... cm²

Probability 1

You will practice how to:

- Understand that complementary events are two events that have a total probability of 1.
- Design and conduct chance experiments or simulations, using small and large numbers of trials. Compare the experimental probabilities with theoretical outcomes.

12.1 Complementary events

Summary of key points

The **complement** of any event A is the event not A, written as A', and can be represented as P(A) + P(A') = 1.

If the probability of an event occurring is p, then the probability of it not occurring is $1 - p$.

Example: The probability that a customer at a pizza restaurant orders cheese and tomato pizza is $\frac{2}{9}$.

The probability that a customer at the restaurant does not order cheese and tomato pizza is $1 - \frac{2}{9} = \frac{7}{9}$.

The probability that the event happens can be written as P(event happens).

Exercise 1

1 The probability that Josh receives post on a day chosen at random is 0.7. Write down the probability that he does not receive post. ....

2 Marco plants some seeds. The probability that a seed will grow is $\frac{7}{8}$. Write down the probability that a seed will not grow. ....

3 The table shows the probability that different trains will arrive on time at a station. Complete the table.

Train	A	B	C	D
P(on time)	0.8	0.9	0.95	0.2
P(not on time)				

4 The table shows the probability that there will be rain on a day chosen at random in different months of the year.

Month	March	June	September	December
P(rain)	27%	9%	13%	29%

What is the probability that it will not rain on a day chosen at random in each of these months? Give each answer as a percentage.

March   ....

June   ....

September   ....

December   ....

5 The table shows the probability that Nina buys different items when she goes to the supermarket.

Item	Probability
Bananas	$\frac{3}{7}$
Biscuits	$\frac{2}{5}$
Bread	$\frac{9}{11}$
Milk	$\frac{11}{12}$

Find the probability that Nina does not buy each item when she goes to the supermarket.

Bananas   ....

Biscuits   ....

Bread   ....

Milk   ....

6 A bag contains 12 counters. The counters are either red or yellow or blue. Anil picks a counter at random from the bag. The probability that the counter is not red is $\frac{2}{3}$. The probability that the counter is not yellow is $\frac{1}{2}$.

a) Find how many blue counters are in the bag.   ....

b) Explain how you worked out your answer to part **a**.

....

....

Think about

7 **Design a spinner with these three properties.**

- Each section is either red or blue or green.

- The probability that the spinner does not land on a red section is $\frac{3}{4}$.

- The probability that the spinner does not land on a blue section is $\frac{5}{8}$.

12.2 Experimental probability

Summary of key points

Some probabilities can be found exactly because outcomes are equally likely. These are called **theoretical probabilities**.

In some situations, an experiment may be needed to estimate a probability. The **experimental probability** of an event can be found using the formula:

Experimental probability = relative frequency = $\dfrac{\text{number of times the event occurred}}{\text{number of trials}}$

Exercise 2

1 **A teacher gives a bag of marbles to each of four children.**

Tia's bag	Tomas' bag	Sofia's bag	William's bag
5 red 5 blue	2 red 6 blue	2 red 1 blue	6 red 1 blue

Each child does an experiment by picking a ball out of their bag 100 times. They put the ball back into the bag after each pick.

Draw a line to match each set of results to the most likely bag.

Colour	Frequency
Red	83
Blue	17

Tia's bag

Colour	Frequency
Red	55
Blue	45

Tomas' bag

Colour	Frequency
Red	23
Blue	77

Sofia's bag

Colour	Frequency
Red	69
Blue	31

William's bag

2 **A six-sided dice is thrown 300 times.**

Score	1	2	3	4	5	6
Frequency	18	54	76	68	47	37

a) Use the results in the table to estimate the probability of getting a 1 on this dice.

....

b) Write down the theoretical probability of getting a 1 on a fair dice.

....

c) Explain why this dice is not a fair dice.

....

....

3 **Kian throws a four-sided dice 80 times, it lands on four 18 times.**

a) Work out the relative frequency of throwing a four with this dice.

....

b) Write down the theoretical probability of throwing a four with a fair four-sided dice.

....

c) Kian says, 'The dice is biased.'

  i) Explain why he may be wrong.

....

  ii) Explain why he may be right.

....

4 A teacher asks 60 children to think of a whole number between 1 and 5 inclusive.
The numbers chosen by the children are shown in the table.

Number chosen	1	2	3	4	5
Number of children	7	20	15	14	4

Do you think the children are choosing a number randomly?
Explain your answer by considering theoretical and experimental probabilities.

....

....

....

....

....

5 Amy throws a six-sided dice 120 times. The table shows her scores.

Score	1	2	3	4	5	6
Frequency	25	11	32	9	26	17

a) Use Amy's results to work out the relative frequency of getting an **even** number on this dice.
Write your answer as a decimal correct to 2 decimal places.

....

b) Write down the theoretical probability of getting an even number on a fair six-sided dice.

....

c) Explain why this dice isn't fair.

....

....

6 Tyson spins a spinner with 3 equal sectors. His results are shown in the table.

Score	Red	Green	Blue
Frequency	25	20	15

Tyson says, 'The spinner is fair because the probability of green is $\frac{1}{3}$.'

Is Tyson correct? Explain your answer.

....

....

13 Calculations

You will practice how to:

- Understand that brackets, indices (square and cube roots) and operations follow a particular order.
- Use knowledge of the laws of arithmetic and order of operations (including brackets) to simplify calculations containing decimals or fractions.

13.1 Order of operations

Summary of key points

In calculations involving several operations, the order to work them out is:

1. Brackets

2. Indices (powers and roots)

3. Multiplication and division (from left to right)

4. Addition and subtraction (from left to right).

Treat calculations under a square or cube root sign like calculations in brackets.

For example, in $3 \times \sqrt{3 \times 12}$, find 3×12 first:

$3 \times \sqrt{3 \times 12} = 3 \times \sqrt{36} = 3 \times 6 = 18$.

Exercise 1

1 Draw a line to match each calculation with the correct answer.

$(72 - 12) \div (3 \times 4)$	1
$\sqrt{20 - 4} - 2$	2
$24 - 5 \times (7 \times 4 - 24)$	3
$36 \div 4 - 4 \times \sqrt{4}$	4
$45 \div (11 + 2 \times 2)$	5

2 Find:

a) $(40 \div 8 + 3 \times 2) \times 4$ ....

b) $(12 - 3 + 6) \times \sqrt{40 - 5 \times 3}$ ....

c) $1 - \sqrt{40 + 3 \times 20} \div 5$ ....

d) $\dfrac{5 + 17}{7 - 18}$ ....

e) $7 \times (30 - (28 - 7))$ ....

f) $63 \div ((11 - 8) \times 3)$ ....

g) $\sqrt[3]{20 - 28} + 7$ ....

h) $\dfrac{3 \times 10}{\sqrt{9 + 16}}$ ....

3 Here are six calculations.

A	B	C
$7 \times 8 - 4^2$	$(11 - 2 \times 3)^2$	$\sqrt{25} \times 3^2$

D	E	F
$(2 \times 5)^2 \div \sqrt[3]{8}$	$5 \times \sqrt{3 \times 12}$	$\left(\dfrac{18}{10 - 7}\right)^2 - 1$

Find the answer to each calculation and write its letter in the correct place in the table.

Calculation A has been done for you.

Calculation				A		
Answer	25	30	35	40	45	50

4 Thalia has done the calculations below.

Mark her work and correct any wrong answers.

a) $27 \div 3^2 - \sqrt{9} = 3$

b) $(12 - 3^2) \times (2^2 + 1) = 15$

c) $20 \div \sqrt{(13 - 3^2)} \times 2 = 5$

d) $\left(\dfrac{3^2 + 6}{3}\right)^2 = 25$

5 Write a whole number to complete each calculation.

a) $40 - \sqrt{\text{.........} - 4 \times 8} = 38$

b) $(11 + \text{...........}) \div 3 + 6 = 16$

c) $(\text{...........} + 12) \times (17 - 3 \times 5) = 38$

d) $(45 + 9) \div (20 - \text{...........}) = 9$

e) $4 + 2 \times (3 + \sqrt[3]{\text{.........}} \times 4) = 26$

f) $75 - (\text{...........} - 3) \times 10 = -5$

6 Insert the correct operation (+, −, × or ÷) to make each calculation correct.

a) $6 \ldots\ldots 3^3 = 33$

b) $10 + 5 \ldots\ldots 4^2 = 90$

c) $(12 \ldots\ldots 4 + 3)^2 = 36$

d) $\dfrac{4^3}{2\ldots\ldots 3 \times 2} = 8$

7 Write whether each statement is true or false.

a) $\dfrac{12 \times 5 + 20}{11 - 8 + 2} = 16$ $\quad\ldots\ldots$

b) $\left(\dfrac{15}{3}\right)^3 - 5 \times 10 = 1200$ $\quad\ldots\ldots$

8 Insert brackets into each calculation to make true statements.
You may need to use more than one pair of brackets, or none at all.

a) $8 + 64 ÷ 4 × 2 + 1 = 17$

b) $8 + 64 ÷ 4 × 2 + 1 = 41$

c) $8 + 64 ÷ 4 × 2 + 1 = 54$

d) $8 + 64 ÷ 4 × 2 + 1 = 49$

e) $8 + 64 ÷ 4 × 2 + 1 = 10$

f) $8 + 64 ÷ 4 × 2 + 1 = 8$

9 Insert a square root sign to make each calculation correct.

a) $100 - 8^2 + 16 = 40$

b) $100 - 8^2 + 16 = 108$

c) $100 - 8^2 + 16 = -38$

d) $100 - 8^2 + 16 = 22$

Think about

10 Create your own questions like the ones in question 9, with whole number answers.

Summary of key points

You can use the **laws of arithmetic** to rewrite some calculations to make them easier to do, without changing the results. The table shows some methods you can use.

Method	Examples
Changing the order of numbers in addition or multiplication (addition and multiplication are **commutative**)	$0.8 + 0.36 + 0.2 = 0.8 + 0.2 + 0.36 = 1 + 0.36 = 1.36$ $4 \times \frac{5}{9} \times \frac{1}{2} = 4 \times \frac{1}{2} \times \frac{5}{9} = 2 \times \frac{5}{9} = \frac{10}{9}$
Changing the order of calculations by changing the grouping in addition or multiplication (addition and multiplication are **associative**)	$\frac{2}{5} + \left(\frac{3}{5} + \frac{5}{6} \right) = \left(\frac{2}{5} + \frac{3}{5} \right) + \frac{5}{6} = 1 + \frac{5}{6} = 1\frac{5}{6}$ $(2.8 \times 2.5) \times 4 = 2.8 \times (2.5 \times 4) = 2.8 \times 10 = 28$
Using the distributive law to rewrite a multiplication by partitioning a number	$\mathbf{1.1} \times 0.6 = \mathbf{1} \times 0.6 + \mathbf{0.1} \times 0.6 = 0.6 + 0.06 = 0.66$ $3\frac{1}{4} \times 12 = 3 \times 12 + \frac{1}{4} \times 12 = 36 + 3 = 39$
Using the distributive law in reverse to rewrite a multiplication	$\mathbf{2.3} \times 3.2 - \mathbf{0.3} \times 3.2 = \mathbf{2} \times 3.2 = 6.4$

Exercise 2

1. Draw a line to match each calculation with its answer.

$(4.6 \times 2) \times 0.5$		4.8
$1.5 \times (0.4 \times 7)$		4.6
$(0.8 \times 12) \times 0.5$		4.9
$3.5 \times 0.7 \times 2$		4.2

2. Complete these calculations.

a) $0.3 \times 1.7 + 0.7 \times 1.7 = \ldots\ldots \times 1.7 = \ldots\ldots$

b) $2.35 \times 2.8 - 0.35 \times 2.8 = 2 \times \ldots\ldots = \ldots\ldots$

c) $\frac{3}{4} \times 11 + 1\frac{1}{4} \times 11 = \ldots\ldots \times \ldots\ldots = \ldots\ldots$

d) 10.2 × 3.6 – .... × 3.6 = 10 × .... = ...

e) $\frac{3}{7}$ × 21 – 21 × $\frac{1}{7}$ = 21 × .... = ...

f) $\frac{1}{9}$ × 38 – 2 × $\frac{1}{9}$ = $\frac{1}{9}$ × .... = ...

3 Find:

a) 0.62 × (0.78 – 0.28)

b) $\left(\frac{5}{8} + \frac{3}{8}\right) \times \frac{3}{10}$

....

....

c) $\frac{9}{10} + \frac{1}{2} \times \frac{2}{5}$

d) $\frac{4}{7} \times \frac{5}{8} + \frac{4}{7} \times \frac{3}{8}$

....

....

e) 1.34 + (0.66 + 0.16)

f) 3.1 × 0.22

....

....

g) $\frac{3}{4} \times \left(\frac{4}{5} + \frac{1}{3}\right)$

h) $\frac{3}{5} + \left(\frac{2}{5} + \frac{9}{10}\right)$

....

....

4 Write the numbers $\frac{3}{4}$, $\frac{1}{4}$, 60 and 4 on the cards to make a correct statement.

Each number should be used only once.

$$\boxed{} \times \boxed{} + \boxed{} \times \boxed{} = \boxed{18}$$

....

Think about

5 Complete this calculation in three different ways.

$$\frac{1}{5} \times \text{........} + \frac{1}{5} \times \text{........} = 6$$

14 Equations and inequalities

You will practice how to:

- Understand that a situation can be represented either in words or as an equation. Move between the two representations and solve the equation (integer or fractional coefficients, unknown on either or both sides).
- Understand that letters can represent open and closed intervals (two terms).

14.1 Solving equations with brackets and fractions

Summary of key points

Example:

Solve $4(3x - 2) = 52$

Expand brackets.	$12x - 8 = 52$
Add 8 to both sides.	$12x = 60$
Divide by 12.	$x = 5$

Exercise 1

1 Solve these equations.

a) $3(x - 1) = 15$

$x = \dots\dots$

b) $5(3x - 4) = 40$

$x = \dots\dots$

c) $49 = 7(1 + 3x)$

$x = \dots\dots$

d) $4(13 - 2x) = 28$

$x = \dots\dots$

2 Solve:

a) $4(2t - 1) + 3(t + 7) = 83$

$t = \dots\dots$

b) $2(3t - 5) + 3(4t - 1) = 59$

$t = \dots\dots$

3 Solve:

a) $\dfrac{1}{3} x = 9$

$x = \dots\dots$

b) $\dfrac{5}{8} y + 13 = 3$

$y = \dots\dots$

4 Connor is trying to solve $7(1 - 2x) = 5$.

He writes:

$$7(1 - 2x) = 5$$
$$7 - 2x = 5$$
$$2x = 2$$
$$x = 1$$

Show Connor how he could have worked out the correct solution to this equation.

....

....

14.2 Solving equations with unknowns on both sides

Summary of key points

To solve an equation with unknowns on both sides, rearrange the equation so that the unknowns appear only on one side.

Remember that what you do to one side, you must also do to the other side.

Example:

Solve $4(x - 2) = 2x + 12$

Start by expanding the brackets.	$4x - 8 = 2x + 12$
Subtract 2x.	$2x - 8 = 12$
Add 8.	$2x = 20$
Divide by 2.	$x = 10$

Exercise 2

1 Solve these equations.

a) $2x + 5 = x + 9$

b) $5x + 3 = x + 7$

x = ...

x = ...

c) $6x - 5 = 3x + 13$

d) $8x - 14 = 2x - 2$

x = ...

x = ...

2 Solve the equations and draw a ring around the odd one out.

$\frac{x}{4} - 5 = 2$

$4x + 1 = 2x + 57$

$4x - 1 = 3x + 29$

x = ...

x = ...

x = ...

3 Solve each equation.

a) $14 - x = x + 2$

b) $y + 7 = 21 - y$

c) $41 - 2n = n + 2$

x = ...

y = ...

n = ...

4 Solve these equations.

a) $5(2k + 7) = 13k + 2$

b) $3(2v - 1) = 39 - v$

k = ...

v = ...

5 Solve these equations.

a) $2(n + 3) + 3(2n - 5) = 4n + 15$

b) $5(r + 10) = 2(5r + 3) + 3r + 4$

n = ...

r = ...

c) $2x - 4 = 5(x - 2) + x + 4$

d) $2(1.75y - 3) = 8 + 3y$

x = ...

y = ...

Think about

6 Write down two different equations that have a solution of *x* = −2.
The equations must contain a bracket.

7 Match each equation with the correct solution.

$2(3x + 7) = 8$

$29 - 11x = x - 7$

$5(2x - 7) = 3(1 - 3x)$

$4\left(\dfrac{1}{2}x - 1\right) = 2 - 6x$

$3(2 - x) = 2x + 26$

$x = 3$

$x = 2$

$x = \dfrac{3}{4}$

$x = -1$

$x = -4$

8 Toby is trying to solve the equation $27 - 7x = 6(x + 11)$.
His working is shown below.

	$27 - 7x = 6(x + 11)$
Expand brackets	$27 - 7x = 6x + 66$
Subtract 27	$7x = 6x + 39$
Subtract 6x	$x = 39$

What mistake did Toby make?

....

Summary of key points

Example:

Arun thinks of a number.

He adds 4, then multiplies the result by 3.

His final answer is 45.

What number does Arun think of?

Let a represent the number Arun thinks of.

Then rewrite the puzzle as the equation. $3(a + 4) = 45$

And solve!

Expand the brackets. $3a + 12 = 45$

Subtract 12. $3a = 33$

Divide by 3. $a = 11$

So Arun thinks of the number 11.

Exercise 3

1 **Anya is n years old.**

When you add 1 to her age and multiply the result by 3, the final answer is 45.

a) Write an equation in terms of n to represent this problem.

....

b) Solve your equation to find the value of n.

$n = $ ....

2 **Jack thinks of a number.**

He multiples it by 4 then subtracts 18 from the result.

His final answer is the same as the number he first thought of.

a) Let m be the number that Jack thought of.

Draw a ring around the equation that matches this number puzzle.

$4m - 18 = 0$ $4m - 18 = m$ $4(m - 18) = 0$ $4(m - 18) = m$

b) Solve the equation to find the value of m.

m = ...

3 Form and solve an equation to find the value of *x*.

a)

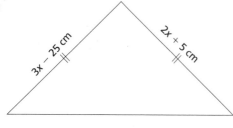

x = ...

b)

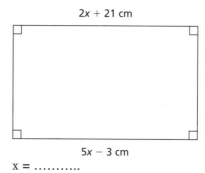

x = ...

4 Three children each form an expression.

Brian's expression   5n − 2

Lola's expression   3(n + 4)

Kim's expression   22 − n

a) Find the value of n such that Brian and Lola's expressions have the same value.

n = ...

b) Kim claims that there is a value of n such that makes all three of their expressions have the same value.

Is she correct?

Kim is correct. ☐   Kim is not correct. ☐

Show how you worked out your answer.

....

....

5 In each part, form and solve an equation to find the value of *a*.

a)

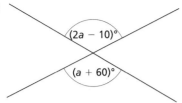

b)

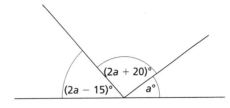

a = ...

a = ...

6 The diagram shows a rectangle.

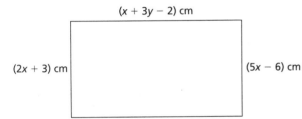

(x + 3y − 2) cm

(2x + 3) cm

(5x − 6) cm

The perimeter of the rectangle is 44 cm.
Find the values of *x* and *y*.

x = ...

y = ...

Think about

7 Make up your own angle problem. Ask a partner to solve it.

8 These two rectangles have the same area. Find the value of *n*.

All lengths are in centimetres.

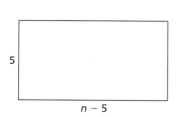

5

n − 5

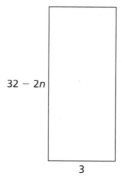

32 − 2*n*

3

n = ...

9 Cakes are sold in small packs and large packs.

A small pack contains *n* cakes.

A large pack contains 9 more cakes than a small pack.

2 large packs contain the same number of cakes as 5 small packs.

Kai buys 2 small packs and 1 large pack.

How many cakes does he buy altogether?

....

Summary of key points

Inequalities can be shown on a number line.

An open circle ○ is used to show < (less than) and > (greater than).

A closed circle ● is used to show ≤ (less than or equal to) and ≥ (greater than or equal to).

 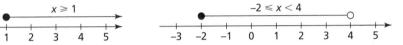

Exercise 4

1 Circle the correct inequalities.

$\pi \geq 3$ $\pi < 3$ $3 < \pi < 4$ $\pi \geq 4$

2 *m* and *n* are two numbers.

Write each statement about *m* and *n* as an inequality.

a) m is greater than n.   m .... n

b) n is a negative number.   n 0

c) m is not a negative number.   m 0

3 a) Write down the smallest whole number that satisfies the inequality n > 5.

....

b) Write down the largest whole number that satisfies the inequality n ≤ 4.

....

4 For each part, write down all the integers that satisfy the inequality.

a) $-3 < x < 2$ ....

b) $-1 \leq x < 3$ ....

c) $1 < x \leq 6$ ....

d) $4 \leq x \leq 6$ ....

5 Match each inequality with the corresponding number line.

$0 < x < 6$

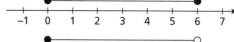

$0 \le x \le 6$

$0 < x \le 6$

$0 \le x < 6$

6 Complete each inequality to match the interval shown on the number line.

a)

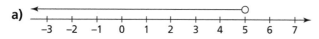

n ....

b)

n ....

c)

.... n ....

7 Insert the correct inequality sign to complete each statement.

a) If $n > 5$, then $3n$ 15

b) If $n < -2$, then $5 - n$ 7

c) If $n > 0$, then $-n$ 0

d) If $n \le -4$, then n^2 16

15 Midpoints

You will practice how to:

- Use knowledge of coordinates to find the midpoint of a line segment.

15.1 Midpoint of a line segment

Summary of key points

A **line segment** is a section of a line. It has a two end points.

The **midpoint** of a line segment is the point on it that is the same distance from both end points. It is halfway along the line segment.

> The x-coordinate of the midpoint is the mean of the x-coordinates of the endpoints.
>
> The y-coordinate of the midpoint is the mean of the y-coordinates of the endpoints.

Exercise 1

1 Mark the midpoint of each line segment with a cross and write down its coordinates.

a) A(0, 0) B(8, 6)

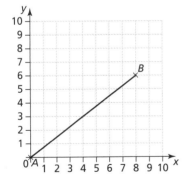

Midpoint (.........., ...)

b) A(1, 7) B(9, 3)

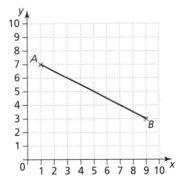

Midpoint (.........., ...)

c) A(2, 10)   B(7, 4)

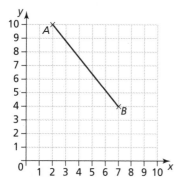

Midpoint (.........., ...)

d) A(1, 2)   B(7, 9)

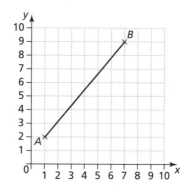

Midpoint (.........., ...)

2 Write down the coordinates of the midpoint *M* of each line segment *AB*.

a) A(0, 4)   B(6, 0)   M(.........., ...)

b) A(1, 7)   B(11, 3)   M(.........., ...)

c) A(2, 12)   B(6, 4)   M(.........., ...)

d) A(9, 4)   B(1, 7)   M(.........., ...)

e) A(−1, 10)   B(7, 0)   M(.........., ...)

f) A(4, −6)   B(8, 4)   M(.........., ...)

g) A(−13, −5)   B(7, 9)   M(.........., ...)

h) A(9, −4)   B(−4, 7)   M(.........., ...)

3 Complete the table.

Coordinates of *A*	Coordinates of *B*	Coordinates of midpoint of *AB*
(11, −4)	(7, 1)	(.........., ...)
(5, −3)	(12, −11)	(.........., ...)
(−1, −6)	(8, −14)	(.........., ...)
(.........., ...)	(7, 9)	(4, 8)
(6, 3)	(.........., ...)	(6, 7)
(.........., −5)	(4, ...)	(8, −8)
(.........., ...)	(−7, −6)	(−4.5, 2)
(.........., 15)	(7, ...)	(5, 6.5)

4 In the triangle *FGH*, *K* is the midpoint of *GH*.

K is a point on *x*-axis such that *FK* is parallel to the *y*-axis.

Work out the coordinates of *H*.

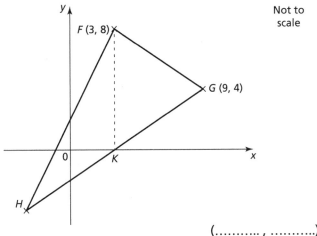

Not to scale

(............ , ...)

5 The diagram shows a sketch of a quadrilateral *PQRS*.

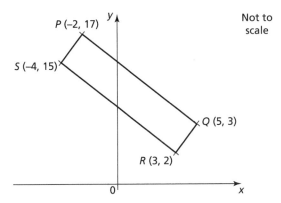

Not to scale

a) Find the coordinates of the midpoint of PR.   (............ , ...)

b) Find the coordinates of the midpoint of QS.   (............ , ...)

c) Jamila says that the quadrilateral is a rectangle.
Explain why Jamila is wrong.

....

....

6 *ABCD* is a parallelogram where:

A is a point with coordinates (1, 2).
B is a point with coordinates (4, 5).
C is a point with coordinates (8, 3).

Show that the diagonals of the parallelogram bisect each other.

....

....

....

....

Think about

7 Draw different quadrilaterals on a coordinate grid. Find the coordinates of the midpoint of each diagonal. For which types of quadrilateral are the coordinates of the two midpoints equal?

16 Fractions, decimals and percentages

You will practice how to:

- Recognise fractions that are equivalent to recurring decimals.
- Understand the relative size of quantities to compare and order decimals and fractions (positive and negative), using the symbols =, ≠, >, <, ≤ and ≥.

16.1 Recurring decimals

Summary of key points

To change a fraction to a decimal, divide the top number (the numerator) by the bottom number (the denominator).

Example: $\frac{2}{9} = 2 \div 9$

$$
\begin{array}{r}
0.\ 2\ \ 2\ \ 2\ ... \\
9\ \overline{\smash{\big)}\ 2\ .\ ^2 0\ \ ^2 0\ \ ^2 0}
\end{array}
$$

So $\frac{2}{9}$ is equivalent to a recurring decimal.

> A **terminating decimal** has a fixed number of decimal places. It does not go on for ever.
>
> The digits after the decimal place in a **recurring decimal** go on forever. A group of digits repeats.

$\frac{2}{9} = 0.\dot{2}$

The dot above the 2 shows that this digit repeats.

Exercise 1

2–3

1. Use a calculator to convert each fraction to a recurring decimal. Fill in the table. The first row has been done for you.

Fraction	Recurring decimal in dot notation	Recurring decimal rounded to 3 decimal places
$\frac{2}{3}$	$0.\dot{6}$	0.667
$\frac{3}{11}$		
$\frac{1}{96}$		
$\frac{22}{27}$		
$\frac{22}{41}$		

2 Work out the missing digits to complete these conversions.

a) $\frac{7}{30}$ = ...

$$\begin{array}{r} 0.2?3\ 3... \\ 30\ \overline{)7.\ ^70\ ^{10}0\ ^{10}0\ ^{10}0} \end{array}$$

b) $\frac{5}{18}$ = ...

$$\begin{array}{r} 0.?\ 7\ ?... \\ 18\ \overline{)5.0^{14}\ 0^7 0^?} \end{array}$$

3 Use a written division method to convert each fraction to a recurring decimal. Use dot notation in your answers.

a) $\frac{2}{9}$

b) $\frac{5}{6}$

c) $\frac{11}{18}$

....

....

....

d) $\frac{5}{12}$

e) $\frac{8}{27}$

f) $\frac{9}{44}$

....

....

....

4 Use a calculator to explore the decimal equivalents of these fractions.

$\frac{17}{99}$ $\frac{23}{99}$ $\frac{41}{99}$ $\frac{50}{99}$

Describe any patterns you can find.

....

....

....

Think about

5 Write the decimal equivalent of $\frac{17}{100}$.

Compare it with the decimal equivalent of $\frac{17}{99}$. Which decimal is greater?

....

---

6 Explain how can you see which is greater, $\frac{17}{100}$ or $\frac{17}{99}$, by looking only at the fractions.

....

16.2 Comparing fractions and decimals

Summary of key points

$x \leq -0.5$ means x is **less than or equal to** -0.5.	
$x \geq 0.6$ means x is **greater than or equal to** 0.6.	
$-0.5 \leq x \leq 0.6$ means the range of x is -0.5 to 0.6 **inclusive**. x can equal -0.5 or 0.6.	$-0.5 < x < 0.6$ means the range of x is -0.5 to 0.6 **exclusive**. x cannot equal -0.5 or 0.6.

Exercise 2

1 $-3.6 \leq x \leq -2.5$

Write:

a) the highest possible value of x ....

b) the lowest possible value of x ....

c) a possible integer value of x ....

d) a possible value of x with 1 decimal place ....

e) a possible value of x with 2 decimal places ....

2 $-0.78 \leq x < -\frac{1}{2}$

Draw a ring around the possible values of x in the list below.

$-\frac{2}{3}$ -0.78 $-\frac{1}{3}$ -0.42 $-\frac{1}{2}$ -0.68 -0.8

3 Omar says, 'If $-0.41 \leq x < -0.36$, then the highest possible value of x is -0.37.' Is he correct? Explain your answer.

....

....

4 Write $\neq$ or $=$ to make correct statements.

a) $\frac{1}{3}$ .... 0.3 b) -0.18 .... $-\frac{9}{50}$ c) $\frac{3}{4}$ .... $\frac{75}{1000}$ d) -0.9 .... $-\frac{1}{9}$

5 Write <, > or = to make correct statements.
For some pairs, it may help to rewrite the fractions with the same denominator or the same numerator.

a) $\dfrac{7}{11}$ .... $\dfrac{7}{8}$ b) $-\dfrac{7}{11}$ .... $-\dfrac{7}{8}$ c) $\dfrac{7}{12}$ .... $\dfrac{5}{12}$ d) $-\dfrac{7}{12}$ .... $-\dfrac{5}{12}$

e) $-\dfrac{2}{5}$ .... $-\dfrac{7}{15}$ f) $\dfrac{4}{17}$ .... $\dfrac{8}{31}$ g) $-\dfrac{3}{16}$ .... $-\dfrac{9}{48}$ h) $\dfrac{4}{37}$ .... $-\dfrac{1}{9}$

6 Write <, > or = to make correct statements.

a) 0.509 .... 0.59

b) −0.8 .... −0.81

c) −0.65 .... − 0.650

d) −0.101 .... −0.11

7 Decide whether each statement is true or false.

	True	False
$-0.7 \le -\dfrac{3}{4}$	☐	☐
$\dfrac{1}{9} \ne 0.111$	☐	☐
$0.3 > \dfrac{1}{3}$	☐	☐
$-\dfrac{27}{20} \le -1.3$	☐	☐

16.3 Comparing fractions by converting to percentages

Summary of key points

If you are comparing fractions, it is sometimes convenient to convert them to percentages.

For example: Which of these proportions is greatest?

$\dfrac{7}{40}$ $\dfrac{4}{25}$ $\dfrac{3}{20}$

Convert them all to percentages:

$\dfrac{7}{40} = \dfrac{175}{1000} = 17.5\%$ $\dfrac{4}{25} = \dfrac{16}{100} = 16\%$ $\dfrac{3}{20} = \dfrac{15}{100} = 15\%$

The greatest proportion is $\dfrac{7}{40}$.

You can also convert to percentages using a calculator (by dividing the numerator by the denominator and then multiplying by 100).

1 Complete the table. Write all fractions in their simplest form.

First number	Second number	First number as a fraction of second	First number as percentage of second
16	80		
33	60		
85	125		
1260	3500		

2 Work out the proportions in A–F. Then write each proportion in the correct column of the table.

A
145 as a percentage of 435
....

B
56 as a percentage of 224
....

C
126 as a percentage of 420
....

D
243 as a percentage of 810
....

E
630 as a percentage of 1800
....

F
3.75 as a percentage of 12.5
....

25%	30%	$33\frac{1}{3}$%	35%

3 Bindesh exercised for 1 hour. He ran for 39 minutes and walked for the rest of the time. He says, 'I ran for 39% of the time!'

Is Bindesh correct? Explain your answer.

....

....

4 Tick the higher mark in each pair.

23%	or	15 out of 60
74%	or	14 out of 20
88%	or	273 out of 300
60%	or	48 out of 75

5 Draw a ring around the highest proportion in each set.

a)   49%   0.44 $\frac{2}{5}$

b)   24%   0.19 $\frac{1}{4}$

c)   0.3   140 out of 420   31%

d) $\frac{19}{25}$ 0.8   630 out of 840

6 Team A has won **13 out of 20** matches.

Team B has won **15 out of 25** matches.

Which team has won a higher percentage of their matches?   ....

7 Phil and Pratima are plumbers.

They ask their customers if they are happy with the work done.

Phil	Pratima
51 out of 60 customers are happy.	132 out of 150 customers are happy.

Which plumber has the higher percentage of happy customers?   ....

8 A farmer records how many apples he picks from three trees and how many of them are small.

Tree A	Tree B	Tree C
13 out of 115 apples are small.	16 out of 134 apples are small	7 out of 79 apples are small.

Work out which tree has the lowest proportion of small apples.

....

....

....

9 The table shows the total areas of three cities and the areas of open space in those cities.

City	Total area (km²)	Area of open space (km²)
City A	892	128
City B	719	115
City C	126	23

The mayor of City A claims that his city has the greatest proportion of open space.

Is the mayor correct?   Yes ☐   No ☐

Show how you worked out your answer.

....

....

....

Presenting and interpreting data 2

You will practice how to:

- Record, organise and represent categorical, discrete and continuous data. Choose and explain which representation to use in a given situation:
  - o pie charts
  - o line graphs and time series graphs
  - o scatter graphs
  - o infographics.
- Interpret data, identifying patterns, trends and relationships, within and between data sets, to answer statistical questions. Discuss conclusions, considering the sources of variation, including sampling, and check predictions.

17.1 Pie charts

Summary of key points

A **pie chart** is a circular diagram. Each category of data is represented as a slice (or sector) of the circle.

Exercise 1

1 A sample of 80 people were asked their favourite swimming stroke.

Complete the table and draw a pie chart to show the results.

Stroke	Frequency	Angle
Front crawl	38	$\frac{38}{80} \times 360 = 171°$
Back crawl	30	
Breast stroke	12	

Favourite swimming strokes

Favourite swimming strokes

2 The pie chart shows the colours of sweets in a packet. There are 600 sweets in the packet.

a) What is the size of the angle for green sweets? ....

Colours of sweets

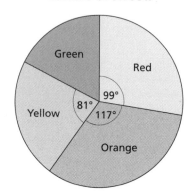

b) What fraction of sweets are yellow? Give your answer in its simplest form.

....

c) Write down the modal colour of sweet. ....

d) How many red sweets are in the packet? ....

3 The table shows what a farmer makes with the milk her farm produces.

What is made	Cheese	Butter	Cream
Percentage of milk	47%	37%	16%

Complete the table and then draw a pie chart to show the information.

What is made	Angle
Cheese	
Butter	
Cream	

4 A postal worker records the size of 250 letters he delivers.

a) Calculate how many large letters he delivers.

....

b) The postal worker says, 'I delivered 40 more small letters than medium letters.'

Is he correct? Show how you worked out your answer.

....

....

....

....

Size of letters

Summary of key points

Infographics show statistical information in a clear and appealing way. They are designed to help illustrate the main features of the data easily.

Exercise 2

1 **The infographic shows sales in a fruit and vegetable shop one day.**

Sales in fruit and vegetable shop

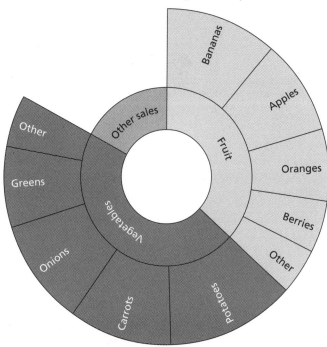

a) Write down which type of vegetable had the highest sales. ....

b) Compare the sales of fruit and vegetables.

    ....

    ....

c) Seema concludes that $\frac{1}{5}$ of all **fruit** sales were bananas.

   Give a reason why she is not correct.

    ....

    ....

d) Write down the two types of vegetables that had equal sales.

.... and ....

2 Real data question **The map shows countries in Southern Africa.**

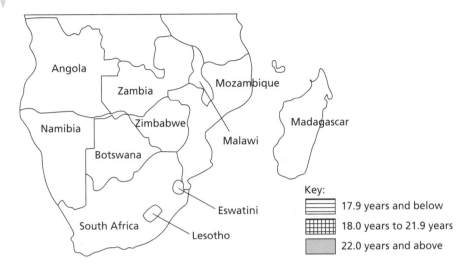

The table shows the median age of people in these countries.

Country	Median age
Angola	16.2 years
Zambia	16.5 years
Mozambique	17.3 years
Zimbabwe	19.2 years
Malawi	17.1 years
Botswana	22.3 years
Namibia	20.9 years
South Africa	25.7 years
Madagascar	18.3 years
Lesotho	20.5 years
Eswatini	19.8 years

Source: WHO W orld Health Statistics

Shade the countries on the map to show this information. Use the key to help you.

3 The diagram shows the sales of books at a bookshop one week.

Books sold one week

				Key:
Fiction		Non-fiction		☐ Fiction
	Children's, 80			☐ Non-fiction
Action, 123		Cooking, 61	History, 56	
	Sci Fi, 67		Nature, 25	
Romantic, 98	Other, 32	Hobbies, 38	Other, 20	

a) Write down the number of Cooking books sold that week. ....

b) Which was the most common type of Fiction book sold? ....

c) Compare the number of History books sold with the number of Nature books sold.

    ....

d) Find the total number of Fiction books sold and the total number of Non-fiction books sold.

    Fiction ....   Non-fiction ....

e) Toby says that 38% of the Non-fiction books were about Hobbies.

    Is he correct? Give a reason for your answer.

    ....

    ....

f) 25% of the Fiction books sold were Hardback.
40% of the Children's books sold were Hardback.

Complete this table to show sales of **Fiction** books.

	Number of Hardback books sold	Number of books sold that are *not* Hardback
Number of Children's books sold		
Number of books sold that are *not* Children's books		

Think about

4 Compare the table in question 3 with a waffle diagram.

Write down how they are similar and how they are different.

17.3 Trends and relationships

Summary of key points

A **time series graph** is a particular type of line graph where time is plotted along the horizontal axis. A time series graph can show **trends** over time – data may show an increasing trend or a decreasing trend.

A **scatter graph** shows the relationship between two variables. A **line of best fit** can be used to estimate the value of one variable when you know the value of the other. Two sets of data can be shown on a single scatter graph – different types of plotting points should be used for each set of data. Remember to include a key.

Exercise 3

1 The diagram shows the population of three countries between 1970 and 2015.

a) Estimate from the graph the population of Ethiopia in 2015.

.... million

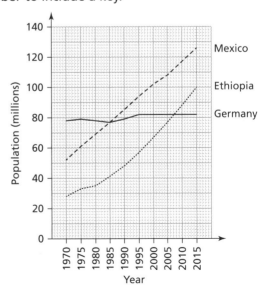

b) Estimate the difference in the populations of Germany and Mexico in 1970.

.... million

c) In which year were the populations of Mexico and Germany approximately equal?

....

2 **The diagram shows the number of tins of soup a factory produced from 2012 to 2017.**

a) How many tins of soup were produced in 2016? .... million

b) In which year did the factory produce 2.4 million tins of soup? ....

c) Between which two years did production decrease?

.... and ....

d) Describe the trend in the number of tins of soup produced.

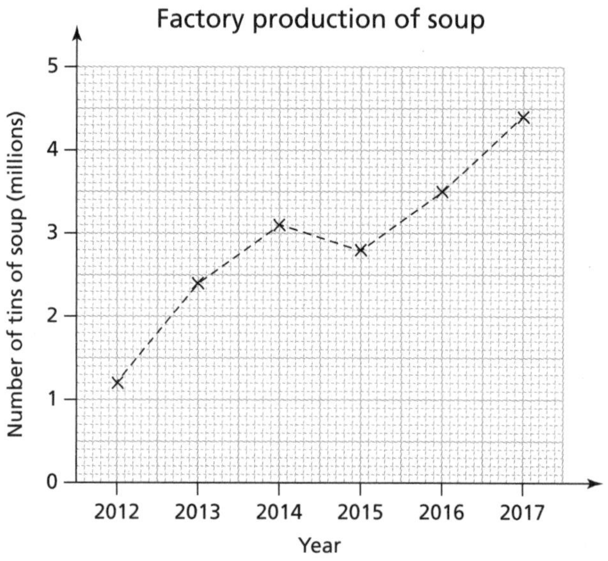

Factory production of soup

....

e) The factory manager claims that the factory produced more than four times as many tins of soup in 2017 than in 2012.

Is he correct?

Yes ☐   ☐ No

Explain your answer.

....

....

3 The time series graph shows the number of laptops a shop sells in each quarter between 2016 and 2018.

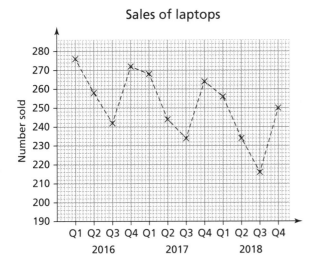

Sales of laptops

a) Write down the number of laptops sold in Q1 2017. ....

b) In which quarter does the shop sell the fewest laptops? ....

c) Describe the trend in the number of laptops sold.

   ....

d) Predict the number of laptops the shop sold in Q1 2019.

   Show how you worked out your prediction.

   ....

4 The tables shows the ages and heights of 7 boys and 7 girls.

Boys:

Age (years)	8	11	13	15	12	9	14
Height (cm)	120	148	154	176	153	137	152

Girls:

Age (years)	10	15	12	14	9	10	8
Height (cm)	129	165	145	158	130	135	117

a) Complete this scatter graph to show this information. Use different plotting symbols for boys and girls. Remember to complete the key.

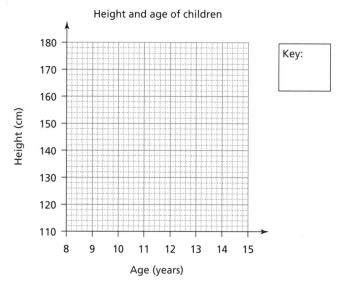

Height and age of children

Key:

b) Describe the relationship between age and height.

....

....

c) Compare the heights of the boys and the girls.

....

....

5 Ada has an oak tree in her garden. Her oak tree produces acorns.
She collects data about the average temperature each spring and the average mass of the acorns on her tree.
This information is shown on the scatter graph.

a) Ada concludes that acorns are heavier when the spring is warmer.

Comment on Ada's conclusion.

....

....

b) Draw a line of best fit on the scatter graph.

c) The following spring the average temperature was 11.8 °C.

Use your line of best fit to estimate the average mass of an acorn on the tree.

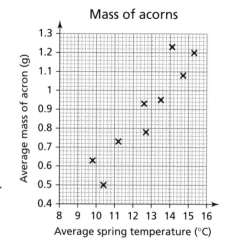

Mass of acorns

.... g

6 The scatter graph shows the value (in $) of small cars of different ages (in years).

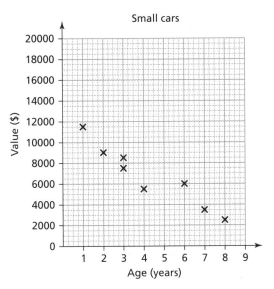

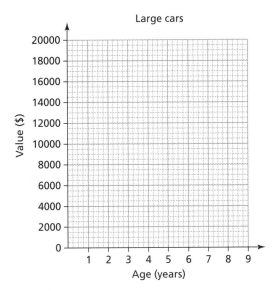

a) The value of large cars is shown in the table.
   Plot the data on the 'Large cars' scatter graph above.

Age (years)	1	1	2	3	4	4	6	7	8
Value ($)	19500	17000	15500	16000	13500	13000	8000	5500	4000

b) Add a line of best fit to both scatter graphs.

c) Predict the value of a 5-year-old small car and a 5-year-old large car.
   Compare these values.

Small car $...............   Large car $.................

Comparison..

....

Transformations

18

You will practice how to:

- Translate points and 2D shapes using vectors, recognising that the image is congruent to the object after a translation.
- Reflect 2D shapes and points in a given mirror line on or parallel to the x- or y-axis, or y = ±x on coordinate grids. Identify a reflection and its mirror line.
- Understand that the centre of rotation, direction of rotation and angle are needed to identify and perform rotations.
- Enlarge 2D shapes, from a centre of enlargement (outside or on the shape) with a positive integer scale factor. Identify an enlargement and scale factor.

18.1 Translations

Summary of key points

We describe translations by using **vectors**. A vector is represented by the combination of a horizontal movement and a vertical movement. The image of a shape after a translation is congruent to the object.

Exercise 1

1 Use vectors to describe these translations.

a) A to B

....

b) A to C

....

c) A to D

....

d) B to C

....

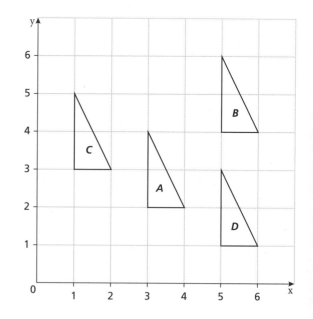

2 Draw the image of triangle *A* after each transformation.

a) Translation with vector $\begin{pmatrix} 4 \\ 2 \end{pmatrix}$.

Label the image P.

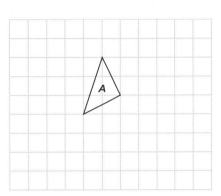

b) Translation with vector $\begin{pmatrix} -3 \\ 2 \end{pmatrix}$.

and label the image Q.

c) Translation with vector $\begin{pmatrix} 4 \\ -3 \end{pmatrix}$.

Label the image R.

3 Match each pair of points to the matching translation.

(3, 4) moves to (5, 1)	$\begin{pmatrix} -2 \\ 3 \end{pmatrix}$
(−3, 4) moves to (−1, 7)	$\begin{pmatrix} 2 \\ -3 \end{pmatrix}$
(3, −4) moves to (1, −1)	$\begin{pmatrix} -2 \\ -3 \end{pmatrix}$
(−3, −4) moves to (−5, −7)	$\begin{pmatrix} 2 \\ 3 \end{pmatrix}$

4 Write down the image of the point (5, −2):

a) after a translation of $\begin{pmatrix} -2 \\ 4 \end{pmatrix}$

....

b) after a translation of $\begin{pmatrix} 2 \\ -3 \end{pmatrix}$.

....

5 The image of the point (1, –3) under a translation is (–4, 4).
Write down the coordinates of the image of (3, –2) under the same translation.

....

6 The vector translating shape *A* to shape *B* is $\begin{pmatrix} 4 \\ -2 \end{pmatrix}$.

Decide whether each statement is true or false.

		True	False
a) The image B is congruent to A.		☐	☐
b) The image of the vertex (2, 1) is (6, –3).		☐	☐
c) The vector translating B to A is $\begin{pmatrix} -4 \\ 2 \end{pmatrix}$.		☐	☐

7 *ABCD* is a kite with vertices at coordinates *A*(1, 0), *B*(3, –1) and *D*(3, 5).
The kite is translated so that point *A* moves to (–8, –3).

a) Write down the vector of the translation.

....

b) Find the coordinates of where point C has been translated to.

(.........., ...)

18.2 Reflections

Summary of key points

The diagram shows the reflection of triangle A in the line y = x.

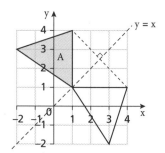

1 Draw the image of quadrilateral *Q* under each transformation.

a) Reflect Q in the line y = 1.
Label the image A.

b) Reflect Q in the line x = 2
Label the image B.

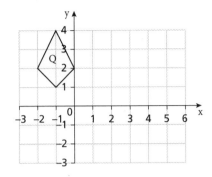

2 Draw the image of shape *L* under each transformation.

a) Reflect L in the line y = −1.
Label the image M.

b) Reflect L in the line x = −1.
Label the image N.

c) Reflect L in the line y = x.
Label the image P.

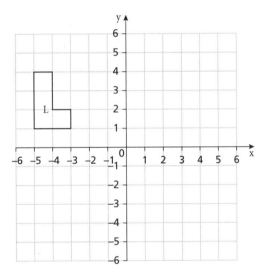

3 Draw the image of triangle *A* under each transformation.

a) Reflect A in the line y = 2.
Label the image M.

b) Reflect M in the y-axis.
Label the image N.

c) Reflect A in the line y = −x.
Label the image P.

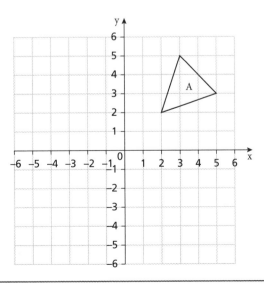

4 The image of a point A(–4, –4) under a reflection is A'(–4, 2).

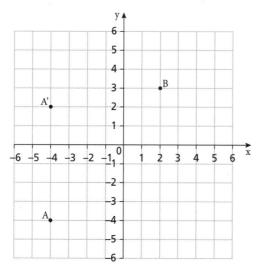

Find the image, B', of the point B(2, 3) under the same reflection.

B' = (………. , ……….)

5 The diagram shows a shape P.

Draw the image of P under a reflection in the line x = 3.

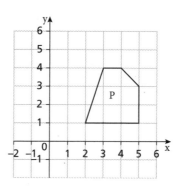

6 The diagram shows six triangles, A, B, C, D, E and F.

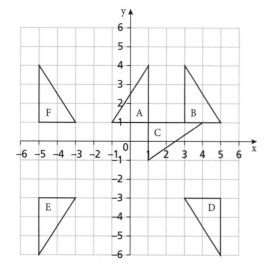

a) State the triangles that are a reflection of triangle A. ....

b) State the triangles that are a reflection of triangle E. ....

c) State the equations of the mirror lines of the following reflections:

i) E to D ....

ii) D to B ....

iii) A to C ....

d) Andrew says, 'Triangle F is a reflection of Triangle B in the y-axis.'
Is Andrew correct? Give a reason for your answer.

....

....

....

18.3 Rotations

Summary of key points

A **rotation** is the turning of a shape. There is an angle and direction of rotation as well as the centre of rotation, around which the shape turns.

Exercise 3

1 Write down the letter of the shape that is the image of trapezium *A* under each transformation.

a) Translation under vector $\begin{pmatrix} 7 \\ 4 \end{pmatrix}$

....

b) Rotation by 180° about point (−2, 1)

....

c) Rotation by 90° anticlockwise about (0, 0)

....

d) Reflection in the line y = −1

....

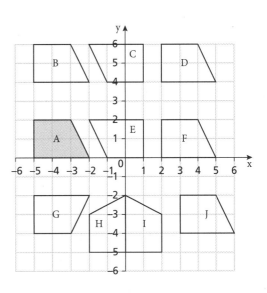

2 Draw the image of quadrilateral *Q* under each transformation.

a) Rotate Q by 90° clockwise, centre (5, 1).
   Label the image A.

b) Rotate Q by 90° anticlockwise, centre (−2, −3).
   Label the image B.

c) Rotate Q by 180°, centre (1, −4).
   Label the image C.

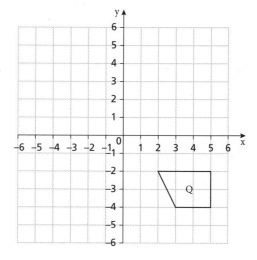

3 Draw the image of parallelogram *P* under each transformation.

a) Rotate P by 90° anticlockwise, centre (2, 2).
   Label the image M.

b) Rotate P by 90° clockwise, centre (3, −1).
   Label the image N.

c) Rotate P by 180°, centre (3, −2).
   Label the image Q.

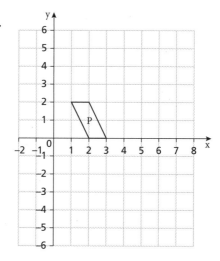

4 Shape *A* has been rotated to the different shapes *B*, *C* and *D*.
Complete the description of each rotation.

a) The rotation of shape A to shape B
   is a rotation of 90° clockwise about
   point (........ , ...).

b) The rotation of shape A to shape C
   is a rotation of ....
   about point (3, 0).

c) The rotation of shape A to shape D
   is a rotation of ....
   about point (−1, ...).

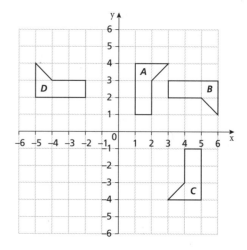

5 Is each statement true or false?

	True	False
a) A 90° clockwise rotation is the same as a 270° anticlockwise rotation.	☐	☐
b) The image of an object under a rotation is congruent to the object.	☐	☐
c) A 180° rotation is the same as a reflection in the line y = x.	☐	☐
d) The 90° clockwise rotation of (4, 5) about (0, 0) is (5, −4).	☐	☐

6 The diagram shows shape *P*. It also shows part of the image of *P* under a single transformation. Complete the image.

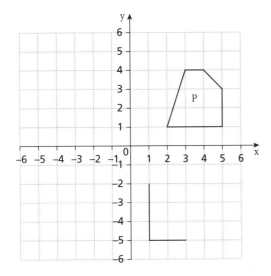

Think about

7 a) Investigate what happens to the point (a, b) when it is rotated by 90° clockwise about the origin.

b) Investigate what happens to the point (a, b) when it is rotated by 90° anticlockwise about the origin.

c) Investigate what happens to the point (a, b) when it is rotated by 180° about the origin.

Summary of key points

An **enlargement** changes the size of a shape to give a similar image.
An enlargement has a centre of enlargement and a scale factor.

Every length of the enlarged shape will be:

Original length × scale factor.

The distance of each image point on the enlargement from the centre of enlargement will be:

Distance of original point from centre of enlargement × scale factor.

For example:

The enlargement shows shape A enlarged to shape B.

The centre of enlargement is (1, 1).
The scale factor is 3.

Each vertex of shape B is 3 times further from the centre of enlargement (1, 1) than the corresponding vertex in shape A.

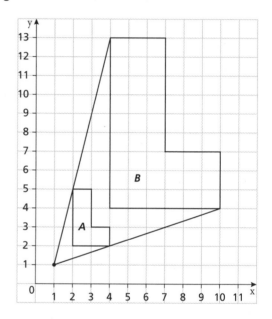

1 Enlarge each shape from the given centre of enlargement by the given scale factor.

a) Centre of enlargement (0, 4), scale factor 2

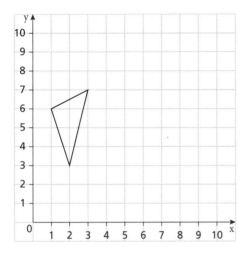

b) Centre of enlargement (1, 4), scale factor 4

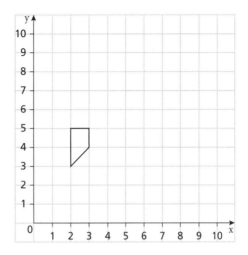

c) Centre of enlargement (3, 3), scale factor 2

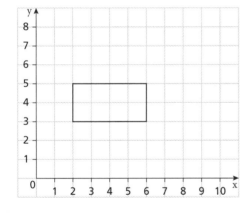

d) Centre of enlargement (10, 7) scale factor 3

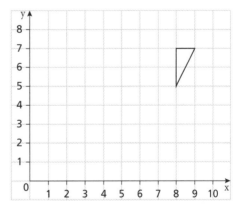

2 **Enlarge each shape from the given centre of enlargement by the given scale factor.**

a) Centre of enlargement (–4, 5), scale factor 2

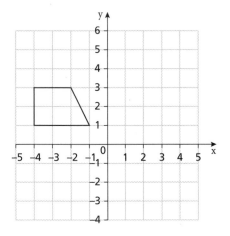

b) Centre of enlargement (4, 3), scale factor 2

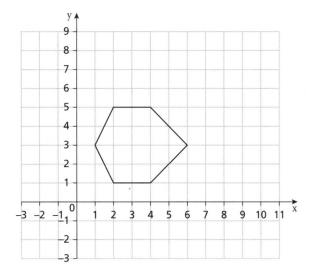

c) Centre of enlargement (–6, –4), scale factor 3

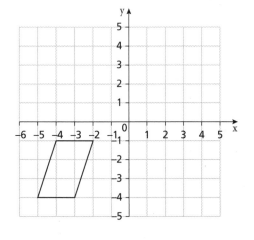

d) Centre of enlargement (11, –3) scale factor 4

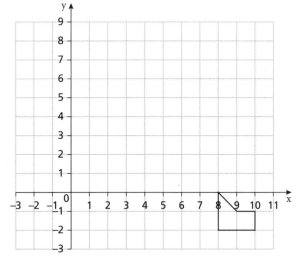

3 Draw the image of the parallelogram *P* under each enlargement.

a) Enlarge P with scale factor 2,
   centre of enlargement (0, 0).
   Label the image Q.

b) Enlarge P with scale factor 3,
   centre of enlargement (1, 3).
   Label the image R.

c) State which shapes are similar to R.

   ....

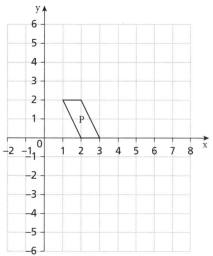

4

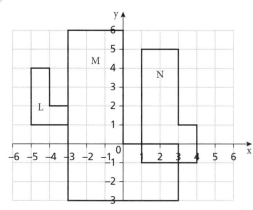

a) Bobby says that shape M is an enlargement of shape L.

   Is Bobby correct?   Yes ☐   No ☐

   If Bobby is correct, state the scale factor of the enlargement.

   ....

b) Jigisha says that shape N is an enlargement of shape L.

   Is Jigisha correct?   Yes ☐   No ☐

   Explain your answer.

   ....

   ....

19 Percentages

You will practice how to:

- Understand percentage increase and decrease, and absolute change.

19.1 Percentage increase and decrease

Summary of key points

There are two methods for calculating a **percentage increase** or **decrease**.

	Method 1 Start by finding the **absolute change**. Then add it to, or subtract it from, the original quantity.	**Method 2** Multiply the original quantity by the correct **multiplier**.
Example: **Increase $300 by 15%.**	First find 15% of $300: 15% of $300 = $45 This is the absolute change. It is a **percentage increase**, so add this to the original amount: $300 + $45 = $345	Use the multiplier 1.15 (because 115% = 1.15): $300 × 1.15 = $345
Example: **Decrease $300 by 8%.**	To decrease $300 by 8%, first find 8% of $300: 1% of $300 = $3 8% of $300 = $24 This is the absolute change. It is a **percentage decrease**, so subtract this from the original amount: $300 − $24 = $276	To decrease $300 by 8%, use the multiplier 0.92 (because 100% − 8% = 92% = 0.92): $300 × 0.92 = $276

Exercise 1

1–3, 6–7

1 Draw a line to match each calculation with its answer.

45% of 1200		140
0.5% of 28 000		240
8% of 5500		340
17% of 2000		440
250% of 96		540

2 Complete the tables to show the multiplier for each percentage change.

Percentage increase	Multiplier
Increase by 42%	………..
Increase by 2.5%	………..
Increase by ………%	1.8
Increase by ………%	4

Percentage decrease	Multiplier
Decrease by 53%	………..
Decrease by 7%	………..
Decrease by ………%	0.8
Decrease by ………%	0.98

3 **a)** Increase 250 by 20%.   **b)** Increase 420 by 35%.

……….   ……….

c) Increase 300 by 110%.   **d)** Decrease 180 by 40%.

……….   ……….

e) Decrease 2000 by 12%.   **f)** Decrease 1500 by 0.1%.

……….   ……….

4 **Gwen uses a calculator to increase 870 by 13%. She types the calculations below.**

$870 \times 0.13 = 113.1$

$870 + 113.1 = 983.1$

Show how to do the calculation in a single step, using a calculator.

……

……

5 Use a calculator to find the answers.

a) Increase $1250 by 26%.

b) Increase 230 km by 232%.

$ ....

.... km

c) Decrease 87.5 g by 8%.

d) Decrease 5400 m by 9.5%.

.... g

.... m

6 A clothes shop reduces all prices by 20% in a sale.

Find the sale price of these items.

a)

$45

b)

$2.40

Original price $45

Original price $2.40

Sale price $...........

Sale price $...........

7 Joselito wants to increase $45 by 200%. His working is below.

For a 200% increase, the multiplier is 2.00.

$45 × 2 = $90

a) Describe the mistake Joselito has made.

....

b) Find the correct answer.

$..........................

Think about

8 Write two questions asking for two different percentage changes, where the absolute change is the same for both.

20 Sequences

You will practice how to:

- Understand term-to-term rules and generate sequences from numerical and spatial patterns (including fractions).
- Understand and describe nth term rules algebraically (in the form $n \pm a$, $a \times n$ or $an \pm b$, where a and b are positive or negative integers or fractions).

20.1 Generating sequences

Summary of key points

Sequences can be continued by identifying the rule that connects the terms.

The **term-to-term rule** for a sequence is 'add 1 and then multiply by 3'.

The first term is 2.

The sequence begins:

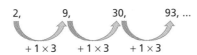

Exercise 1

1 Match each sequence with its term-to-term rule.

Multiply by 2 and then subtract 1	4, 4, 4, ...
Divide by 2 and then add 6	4, 6, 10, ...
Multiply by 3 and then subtract 8	4, 7, 13, ...
Subtract 1 and then multiply by 2	4, 8, 10, ...

2 Write the term-to-term rule for each sequence and use it to find the 8th term.

a) 100, 89, 78, 67, ...   Subtract ....   8th term = ...

b) −16, −13, −10, −7, ...   8th term = ...

c) 15, 11, 7, 3, ...   8th term = ...

d) 6.7, 7.2, 7.7, 8.2, ...   8th term = ...

e) $5\frac{1}{4}$, $6\frac{1}{2}$, $7\frac{3}{4}$, 9, ...   8th term = ...

3 Find the required term in each sequence.

	Term	Term-to-term rule	Term to be found
a)	1st term = 3.4	Subtract 0.3	5th term = ...
b)	1st term = 1.2	Add 0.5	6th term = ...
c)	3rd term = −13	Subtract 3	1st term = ...
d)	6th term = 4.5	Add 0.7	1st term = ...
e)	1st term = 5	Multiply by 2 and then subtract 1	4th term = ...
f)	1st term = −4	Multiply by −1 and then add 3	3rd term = ...
g)	3rd term = 22	Add 1 and then multiply by 2	1st term = ...
h)	4th term 6.4	Multiply by 2 and then add 0.8	1st term = ...

4 Here is a pattern made from squares.

Pattern 1   Pattern 2   Pattern 3

Find how many squares there will be in Pattern 6. ....

5 Here is a pattern made from circles.

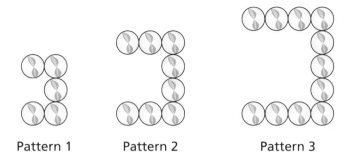

Pattern 1   Pattern 2   Pattern 3

Find how many circles there will be in Pattern 8.

....

6 Here is a pattern made from circular tiles.

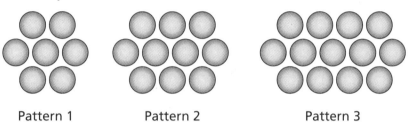

Pattern 1   Pattern 2   Pattern 3

Freda has 60 circular tiles. She wants to make one of the patterns in the sequence.

What is the largest pattern number she can make with her tiles?

....

20.2 nth term rules

Summary of key points

The sequence 4, 8, 12, 16, . . . is the sequence of numbers in the 4 times table.

The position-to-term rule is **multiply by 4**.

position number $\longrightarrow$ $\boxed{\times 4}$ $\longrightarrow$ term

The term in position n is $4n$.

The sequence 7, 11, 15, 19, . . . also increases by 4 each time.

Each term in this sequence is **3 more** than the corresponding number in the 4 times table.

So the position-to-term rule is **multiply by 4 and add 3**.

The expression for the nth term is $4n + 3$.

1 Complete the table.

Position-to-term rule	First four terms of sequence
Multiply by 3 and add 4	.........., ..., ..., ...
Multiply by –4 and add 25	.........., ..., ..., ...
Multiply by 5 and add	8, 13, ..., ...

2 Find the 5th term in the sequence generated by each nth term rule.

a) $2n + 14$ ....   b) $45 - 6n$ ....   c) $1.5n - 3$ ....

3 The position-to-term rule of a sequence is

Multiply by –3 and add 40.

a) Which term in the sequence is equal to 25?

....

b) What is the position of the first negative term in the sequence?

....

4 Complete the table.

	Times table sequence	Related sequence
a)	5, 10, 15, 20, . . . nth term = ...	6, 11, 16, 21, . . . nth term = ...
b)	3, 6, 9, 12, . . . nth term = ...	7, 10, 13, 16, . . . nth term = ...
c)	7, 14, 21, 28, . . . nth term = ...	5, 12, 19, 26, . . . nth term = ...
d)	6, 12, 18, 24, . . . nth term = ...	2, 8, 14, 20, . . . nth term = ...

5 Write down the nth term rule for each sequence.

a) 10, 20, 30, 40, . . .   nth term = ...

b) 9, 17, 25, 33, . . .   nth term = ...

c) 3, 7, 11, 15, . . .   nth term = ...

d) –3, 2, 7, 12, . . .   nth term = ...

6 A sequence begins   11, 20, 29, 38, . . .

   a) Find the nth term of the sequence.   nth term = ...

   b) Find the 20th term of the sequence.   ....

   c) Which term in the sequence is equal to 254?   ....

7 Match each sequence with its rule.

20, 17, 14, 11, ...		$31 - 5n$
26, 21, 16, 11, ...		$27 - 3n$
23, 19, 15, 11, ...		$23 - 3n$
24, 21, 18, 15, ...		$27 - 4n$

8 Write an expression for the *n*th term of each sequence.

   a) 38, 34, 30, 26, . . .   nth term rule = ...

   b) 39, 36, 33, 30, . . .   nth term rule = ...

   c) 60, 49, 38, 27, . . .   nth term rule = ...

9 Sequence A begins   –8, –2, 4, 10, . . .

   The *n*th term of Sequence B is $5n - 1$

   a) Write an expression for the nth term of Sequence A.   ....

   b) Which term in Sequence B is the same as the 18th term in Sequence A?

          ....

10 Adrian makes this pattern using black and white counters.

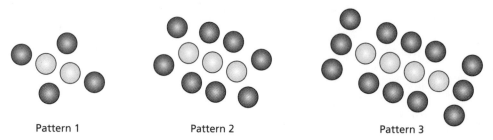

| Pattern 1 | Pattern 2 | Pattern 3 |

Adrian writes this expression for the total number of counters in Pattern *n*.

$$4n + n + 1$$

Are these statements true or false?

	True	False
$4n$ is the expression for the number of black counters in Pattern n.	☐	☐
$n + 1$ is the expression for the number of white counters in Pattern n.	☐	☐
The total number of counters in Pattern 8 is 40.	☐	☐
Adrian's expression simplifies to $6n$.	☐	☐

11 Claudi makes this pattern using square and circular tiles.

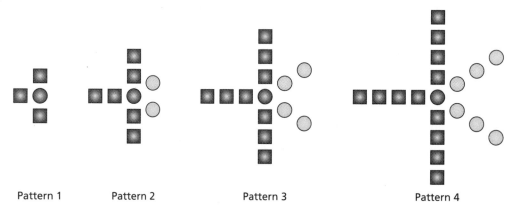

| Pattern 1 | Pattern 2 | Pattern 3 | Pattern 4 |

a) How many black square tiles will there be in the 20th pattern? ....

b) One of the patterns in the sequence has 16 white circular tiles. How many black square tiles are in this pattern?

....

c) Put a ring around the correct expression for the total number of black tiles in Pattern n.

n + 3   3n   3n + 1   4n

d) Put a ring around the **two** correct expressions for the number of white circular tiles in Pattern n.

n + 2   2n   2n − 2   2(n − 1)   2 + n

12 **Write down the nth term rule for each sequence.**

a) 4.6, 5.6, 6.6, 7.6, ...

b) $\frac{5}{8}$, $1\frac{5}{8}$, $2\frac{5}{8}$, $3\frac{5}{8}$, ...

c) 2.1, 4.2, 6.3, 8.4, ...

d) 1.3, 2.1, 2.9, 3.7, ...

e) $5\frac{1}{3}$, $8\frac{1}{3}$, $11\frac{1}{3}$, $14\frac{1}{3}$, ...

f) 6.0, 4.9, 3.8, 2.7, ...

Think about

13 **Make up some of your own arithmetic sequences and find the *n*th term for each one. Try to choose your sequences so that:**

- **the first term of one of the sequences is 10**
- **one of the sequences decreases by 5 each time**
- **the 7th term of one of the sequences is 32**
- **the terms of one of the sequences are negative from the 3rd term onwards.**

Probability 2

You will practice how to:

- Understand that tables, diagrams and lists can be used to identify all mutually exclusive outcomes of combined events (independent events only).
- Understand how to find the theoretical probabilities of equally likely combined events.
- Record, organise and represent categorical, discrete and continuous data.
  o Venn diagrams

21.1 Lists and tree diagrams for combined events

Summary of key points

The **event** of flipping a coin has two **outcomes**: heads or tails.

Two outcomes are **mutually exclusive** if they cannot happen at the same time.

When you work through a problem **systematically** you should follow a logical order to ensure that nothing gets missed.

Successive events occur one after the other or at the same time.

Tree diagrams are used to display combinations of two or more events and are a way to make sure no events are missed out.

Exercise 1

1 Pria picks one of these cards at random.

a) List all possible outcomes for the number on her chosen card.

....

b) List all possible outcomes for the shape on the card she picks.

....

2 Jade chooses a sandwich and a drink from the menu.

Sandwiches | Drinks
Egg | Orange juice
Cheese and tomato | Milk
Tuna | Water

a) Complete the tree diagram to show all the possible combinations.

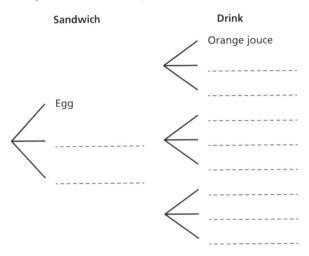

Sandwich

Drink

Orange jouce

Egg

b) Jade chooses a sandwich and a drink at random.

Write down the probability she has a tuna sandwich and water.

....

3 Carla and Frank each choose a numbered ball at random from their own bag.

Carla's bag    Frank's bag

a) Complete the table to show all possible combinations.

Carla's number	Frank's number
2	1
2	2
2	3
2	4

b) Find the probability that the number on Carla's ball is the same as the number on Frank's ball.

....

c) Find the probability that the sum of the numbers on the two balls is 7.

....

4 Angela spins both of these spinners.

Spinner A   Spinner B

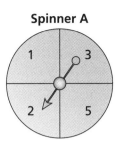 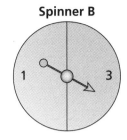

a) Complete the tree diagram to show all the possible combinations.

Spinner A   Spinner B

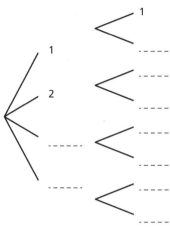

b) Find the probability that the sum of the numbers on the two spinners is less than 5.

....

5 Nina spins a coin three times.

a) List all the possible outcomes.

....

....

b) Nina says that she can get 0, 1, 2 or 3 heads, so the probability of getting exactly 2 heads from three spins must be $\frac{1}{4}$.

Show that Nina is wrong.

....

....

Summary of key points

A **sample space diagram** can show all the possible outcomes when two experiments are performed.

A **sample space** is a list of all the possible outcomes of an experiment.

Example:

A coin is spun and a dice is thrown. The sample space diagram shows the 12 possible outcomes.

	1	2	3	4	5	6
Heads	1, H	2, H	3, H	4, H	5, H	6, H
Tails	1, T	2, T	3, T	4, T	5, T	6, T

The probability of getting Tails and a number greater than 4 is $\frac{2}{12} = \frac{1}{6}$ (5, T or 6, T).

The equally likely outcomes can be sorted into a **Venn diagram**:

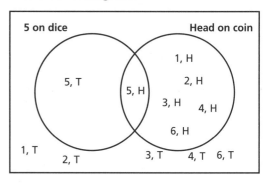

Exercise 2

1 A bag contains three balls: one red, one blue and one green.

A ball is chosen at random and put back in the bag. A second ball is then chosen at random.

a) Complete the tree diagram to show all the possible outcomes.

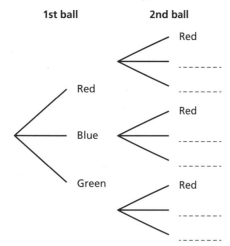

b) Find the probability that the two balls are the same colour.

....

c) Event A is the first ball chosen is a blue ball.

Event B is the second ball chosen is a blue blue.

Complete the Venn diagram to show all nine combinations.

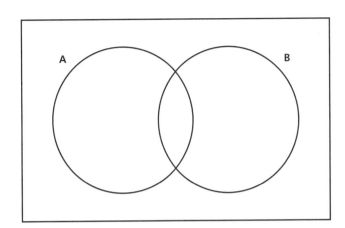

2 Jan has two sets of cards.

Set 1:

Set 2:

She takes one card at random from each set.

The sample space diagram shows the possible outcomes.

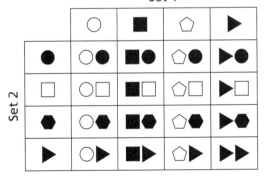

a) Find the probability that Jan picks exactly one triangle.

....

b) Find the probability that Jan picks two black shapes.

....

3 **An ordinary dice is thrown twice and the product of the two scores is found.**

a) Complete the table to show all the possible products.

1st throw

×	1	2	3	4	5	6
1	1	2	3	4		
2						
3						
4			12			
5						30
6						

2nd throw

b) Find the probability that the product of the two scores is an even number.

....

c) Find the probability that the product of the two scores is greater than 21.

....

4 **Grigor has two bags containing coloured marbles. The marbles are either Blue (B), Red (R) or Green (G).**

Bag 1   Bag 2

He takes a marble at random from each bag.

a) Complete the sample space diagram to show the possible outcomes.

Bag 1

	Red	Red	Blue	Green
Red	R, R			
Blue	R, B			

Bag 2

b) Find the probability that Grigor picks at least one red marble. ....

c) Grigor says 'The most likely combination is two red marbles.'

Is Grigor correct?   Yes ☐   No ☐

Explain your reasoning.

....

....

d) Event A is picking a red marble from bag 1.

Event B is both marbles are the same colour.

Write all 16 possible outcomes in the correct position on the Venn diagram.

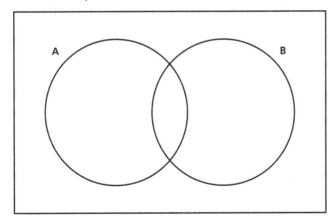

5 Soo-Jung has a fair spinner.

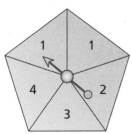

Soo-Jung spins the spinner twice and adds the scores.

a) Draw a sample space diagram to show all the possible totals.

1st spin

+	1	1	2	3	4

2nd spin

b) Find the probability that the total of the two scores is 3. ....

c) Which total score is most likely? ....

d) Find the probability that the total score is more than 4. ....

6 Maxine and Ben each make a fair spinner.

Maxine's spinner has three sides and Ben's spinner has four sides.

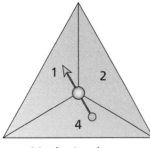

Maxine's spinner

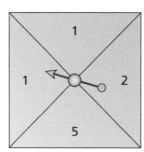

Ben's spinner

They each spin their own spinner twice and add the two scores together.

Ben says that he has a higher probability than Maxine of getting a total score of at least 5.

Is Ben correct?   Yes [ ]   No [ ]

Use sample space diagrams to show your working.

Think about

 Design your own spinner.

Show, on a sample space diagram, all the possible outcomes when your spinner is spun twice and the scores combined. Use your diagram to write some probability statements.

22 Ratio and proportion

You will practice how to:

- Understand and use the relationship between ratio and direct proportion.
- Use knowledge of equivalence to simplify and compare ratios (different units).
- Understand how ratios are used to compare quantities to divide an amount into a given ratio with two or more parts.

22.1 Relating ratio and proportion

Summary of key points

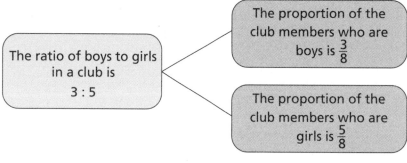

The ratio of boys to girls in a club is

3 : 5

The proportion of the club members who are boys is $\frac{3}{8}$

The proportion of the club members who are girls is $\frac{5}{8}$

Exercise 1

1 **This rectangle is divided into equal-sized squares.**

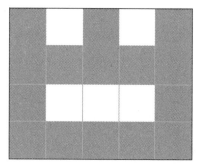

a) Find the ratio of grey to white squares. Write it in its simplest form.

grey : white = ……….. : ………..

b) What proportion of the rectangle is shaded grey? Write this as a fraction in its simplest form.

………………

2 Letitia has five bags containing red and yellow sweets.

Match each ratio to the correct fraction.

Ratio red : yellow | Fraction of sweets that are red

2 : 1	$\frac{1}{8}$
1 : 2	$\frac{1}{3}$
3 : 4	$\frac{3}{8}$
1 : 7	$\frac{2}{3}$
3 : 5	$\frac{3}{7}$

3 Jamila says, 'If I get $\frac{7}{10}$ of some sweets, that means I get 7 sweets.'

Is she correct? Explain your answer.

....

....

4 Emilio and Anders go on holiday. The holiday costs $750. Emilio pays $400.

a) What proportion does Emilio pay? Write this as a fraction in its simplest

form. ....

b) What proportion does Anders pay? Write this as a fraction in its simplest

form. ....

c) Write the ratio of the amount Emilio pays to the amount Anders pays,

in its simplest form. ....

5 Tarif has blue socks and black socks. $\frac{7}{12}$ of his socks are blue.

a) What proportion of his socks are black? ....

b) Write the ratio of blue socks to black socks. ....

6 A shop sells green apples and red apples. The ratio of green apples to red apples is 5 : 2.

a) What fraction of the apples are green? ....

b) What fraction of the apples are red? ....

7 At a restaurant, children can choose pizza or curry for their lunch.

64% of the children choose pizza.

Find the ratio of children choosing pizza to children choosing curry.
Write it in its simplest form.

....

22.2 Simplifying and comparing ratios

Summary of key points

Simplifying ratios

To simplify a ratio, divide or multiply each part by the same number, to get the smallest possible integer values.

Example: Simplify the ratio 2 minutes : 100 seconds Before simplifying, convert the quantities to the same unit. Then divide by 20. 120 seconds : 100 seconds = 120 : 100 $= 6 : 5$ Write the final ratio without the unit.	Example: simplify the ratio $$\frac{1}{3} : \frac{5}{6} : 1\frac{1}{3}$$ Write the fractions with the same denominator. Then multiply by 6. $$\frac{2}{6} : \frac{5}{6} : \frac{8}{6} = 2 : 5 : 8$$

Comparing ratios

The ratio of full-time to part-time staff in company A is 5 : 3.

The ratio of full-time to part-time staff in company B is 8 : 5.

Find which company has the greater proportion of part-time staff.

Method 1	Method 2
Find equivalent ratios that are easier to compare. For example: Company A 5 : 3 = 40 : 24 Company B 8 : 5 = 40 : 25 or, using unitary ratios: Company A 5 : 3 = 1 : 0.6 Company B 8 : 5 = 1 : 0.625 So Company B has the greater proportion of part-time staff.	Compare the proportions of part-time staff. For Company A, this is $\frac{3}{8} = 0.375$ For Company B, this is $\frac{5}{13} = 0.384...$ So Company B has the greater proportion of part-time staff.

Exercise 2

1–8, 10

1 **Write each ratio in its simplest form.**

a) 28 : 12

....

b) 4 : 10 : 6

....

c) 21 : 28 : 63

....

d) 18 : 9 : 45

....

e) 30 : 120 : 180

....

f) 32 : 160 : 48

....

2 **Jurgen has some badges. 18 are squares, 36 are rectangles and 24 are circles.**
Write each ratio in its simplest form.

a) squares : rectangles = ... : ....

b) rectangles : circles = ... : ....

c) square : rectangles : circles = ... : .... : ....

3 **Here are some patterns made from coloured tiles.**
For each pattern, write the ratio in its simplest form.

Pattern	**Ratio** **black tiles : white tiles : grey tiles**
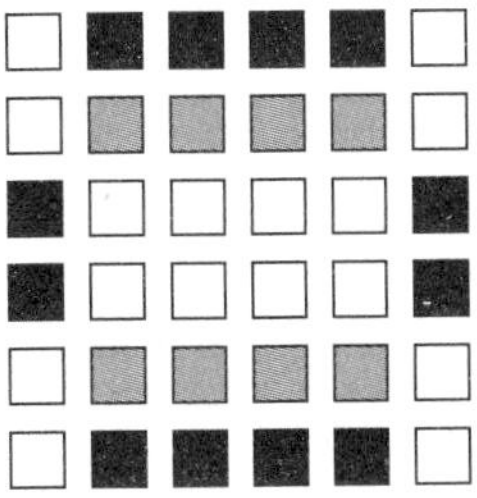	.... : .... : ....
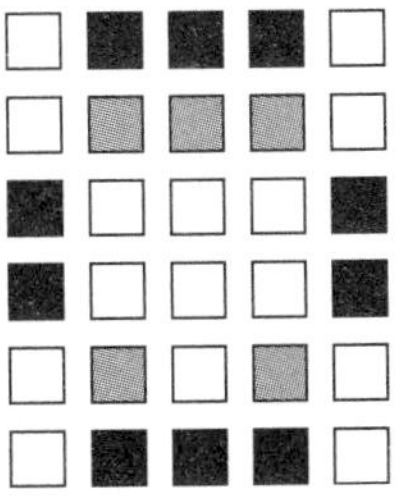	.... : .... : ....
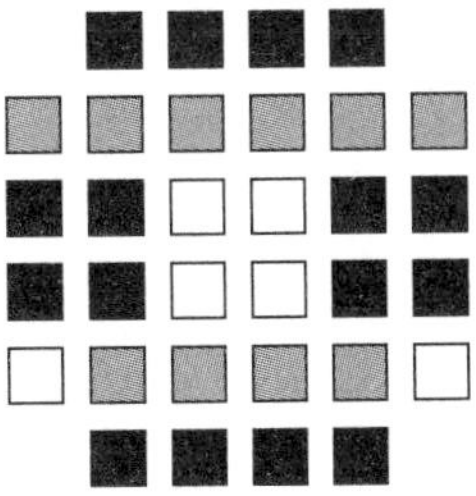	.... : .... : ....

4 Write each ratio in its simplest form.

a) 0.2 : 0.3 : 4

b) 0.5 : 1.5 : 2.5

....

....

c) 0.6 : 0.8 : 1.4

d) $3 : \frac{1}{2} : 2$

....

....

e) $\frac{1}{5} : \frac{3}{5} : \frac{2}{5}$

f) $\frac{2}{3} : \frac{5}{6} : \frac{1}{3}$

....

....

5 Simplify fully:

Remember to change the quantities into the same units, where necessary, before simplifying.

a) 72 km : 45 km

b) 1.4 m : 70 cm

.... : ....

.... : ....

c) 600 g : 1.5 kg

d) 60 mm : 4.5 cm : 15 mm

.... : ....

.... : .... : ....

e) 960 ml : 800 ml : 1.6 litres

f) 4 minutes : 180 seconds : 270 seconds

.... : .... : ....

.... : .... : ....

6 A shop sells bottles of water in three sizes: 250 ml, 500 ml and 1.5 litres. Write the ratio of the sizes of these bottles in its simplest form.

.... : .... : ....

7 Ali has eight bags of sweets containing red sweets and yellow sweets only.

The ratio of red sweets : yellow sweets in Bag A is 1 : 3.
The ratios of red sweets : yellow sweets in each of the other bags are listed below.

1 : 5		2 : 1		2 : 5		4 : 13
	3 : 1		1 : 1		2 : 11	

Draw a ring around the bags that have a greater proportion of yellow sweets than Bag A.

8 Sasha and Eva mix yellow paint and blue paint to make green paint.

Sasha's paint	Eva's paint
litres of yellow : litres of blue	litres of yellow : litres of blue
3 : 7	4 : 9

Whose paint contains the greater proportion of blue paint?
Show your working.

....

9 A cinema showed two films on Tuesday afternoon. Both films were watched by adults and children.

Film 1	Film 2
adults : children = 5 : 8	adults : children = 7 : 11

Show which film was watched by a greater proportion of children.

....

Think about

10 A shade of pink paint is made by mixing red paint and white paint in the ratio 2 : 3.

Find a ratio of red : white paint that will make:
- a darker shade of pink
- a paler shade of pink.

Summary of key points

Example: Share 270 grams in the ratio 2 : 3 : 4.

Number of parts = 2 + 3 + 4 = 9

1 part = 270 g ÷ 9 = 30 g

So, 2 parts = 60 g and 3 parts = 90 g and 4 parts = 120 g.

Exercise 3

1. **Here is a rectangle made of small squares.**

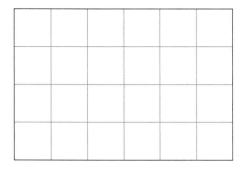

   a) Shade the small squares in black, grey and white so that the ratio of black to grey to white squares is 4 : 3 : 1.

   b) Write the fraction of the rectangle that is grey, in its simplest form.

   ....

2. **Share each amount in the given ratio.**

	Amount	Ratio	Answer
a)	$160	3 : 2	$.......... and $..........
b)	$320	3 : 5	$.......... and $..........
c)	$48	1 : 2 : 3	$.........., $.......... and $..........
d)	$96	3 : 4 : 5	$.........., $.......... and $..........

3 Maxine has 1.2 kg of flour.

She uses some of the flour to make bread, some to make a cake and the rest to make biscuits.

She uses the flour to make each item in the ratio

bread : cake : biscuits = 5 : 2 : 3

Find the amount of flour (in grams) Maxine uses for each item.

Bread .... g   Cake .... g   Biscuits .... g

4 All students at a school choose to study exactly one of Drama, Art or Music.

The ratio of the number of students taking these subjects is:

Drama : Art : Music = 3 : 7 : 2

There are 492 students at the school.

a) Carli says that 260 students study Art and 155 students study Drama.

Without doing any detailed calculations, explain can you tell that Carli must have made a mistake.

....

....

b) Find how many students study Drama.

....

5 In one week, the ratio of Dakarai's hours of sleep, work and free time is 8 : 6 : 7.

a) Find the number of hours he spends working in one week.   ....

b) Find the average number of hours of free time he has in one day. ....

6 A tin contains 60 biscuits.

All the biscuits in the tin are either rectangular or circular.
The ratio of rectangular to circular biscuits is 1 : 2.

All the biscuits are covered with either milk chocolate or dark chocolate.
The ratios are shown below.

Rectangular biscuits
milk chocolate : dark chocolate
3 : 1

Circular biscuits
milk chocolate : dark chocolate
3 : 2

Find how many more of the biscuits are covered in milk chocolate than dark chocolate.

....

23 Relationships and graphs

You will practice how to:

- Use knowledge of coordinate pairs to construct tables of values and plot the graphs of linear functions, where y is given explicitly in terms of x ($y = mx + c$).
- Recognise that equations of the form $y = mx + c$ correspond to straight-line graphs, where m is the gradient and c is the y-intercept (integer values of m).
- Understand that a situation can be represented either in words or as a linear function in two variables (of the form $y = mx + c$), and move between the two representations.
- Read and interpret graphs with more than one component. Explain why they have a specific shape and the significance of intersections of the graphs.

23.1 Plotting graphs of linear functions

Summary of key points

A **linear function** is any function that graphs to a straight line.

For example, $y = 4x - 3$, $y = 2x + 7$ and $y = 5 - \dfrac{x}{2}$ are all linear functions.

To draw the graph of a function, create a table of values, then plot and join the points.

Example: Draw the graph of $y = 2x - 1$.

x	−2	−1	0	1	2
y	−5	−3	−1	1	3

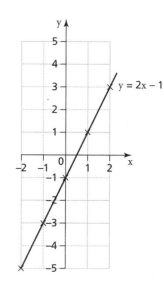

Exercise 1

 a) Complete the table of values for $y = 2x + 4$.

x	−1	0	1	2	3
y				8	

b) Complete the table of values for $y = 3x - 2$.

x	−1	0	1	2	3
y			1		

c) Draw the graphs of $y = 2x + 4$ and $y = 3x - 2$ on the grid.

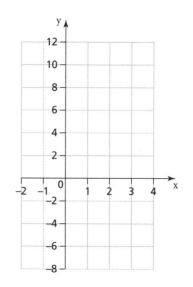

2 a) Complete the table of values for $y = \frac{1}{2}x + 4$.

x	−2	0	2	4
y				6

b) Complete the table of values for $y = 8 - 3x$.

x	−2	0	2	4
y		8		

c) Draw the graphs of $y = \frac{1}{2}x + 4$ and $y = 8 - 3x$ on the grid.

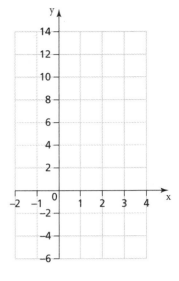

3 a) Draw the graphs of $y = 3x - 4$ and $y = 6 - 2x$ on the grid. Complete the tables of values to help you.

x			
y = 3x − 4			

x			
y = 6 − 2x			

b) Write down the coordinates of the point where the two

lines cross. (..........., ...)

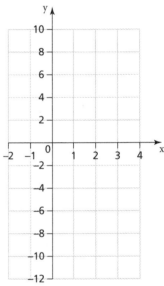

4 A straight line has equation **y = x + 1**.

a) What is the y-intercept? ....
Plot it on the axes.

b) What is the gradient? ....

c) Start at the y-intercept.
Draw a straight line with this gradient.
Extend your line to both edges of the grid.
Label your line.

5 Use the method and grid in question 4 to plot and label these graphs.

a) $y = 2x - 1$ **b)** $y = x + 3$

c) $y = 7 - 2x$ **d)** $y = \frac{1}{2}x + 2$

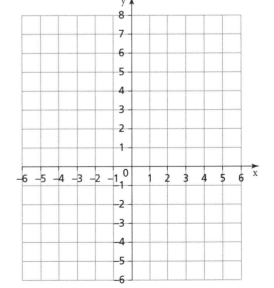

6 Jake draws the graph of $y = 2x - 4$.

He writes:

If $y = 0$, then $x = 2$.
So (2, 0) lies on the line and this is where it crosses the x-axis.
If $x = 0$, then $y = -4$.
So (0, −4) lies on the line and this is where it crosses the y-axis.

Jake joins the points and draws the line.

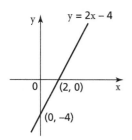

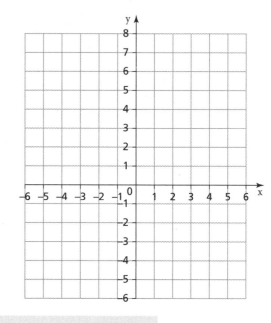

Use Jake's method to draw these lines:

a) $y = 5 - x$

b) $y = 2x - 6$

Think about

7 Joanne says, 'I can't use Jake's method to draw the graph of $y = 3x$.'

Joanne is correct. Explain why.

....

....

....

23.2 Equation of a line

> The **gradient** of a line is a measure of how steep it is.
>
> If the equation of the line has the form $y = mx + c$, m represents the gradient and c represents the y-intercept.

Summary of key points

A formula for the gradient of a line is:

$$\text{gradient} = \frac{\text{increase in } y\text{-coordinate}}{\text{increase in } x\text{-coordinate}}$$

Exercise 2

1 Draw a ring around the equations of lines with the gradient 4.

$y = 4x – 2$ $y = x + 4$ $y = 4$ $y = 5 – 4x$

$y = \dfrac{x}{4} – 4$ $y = 4x$ $x = 4$ $y = 3 – 4x$

2 Choose the correct y-intercept of each straight line from the options in the box below.

–2	–1	1	2	2.5	3	4	5	6

a)   **b)**   **c)**   **d)**

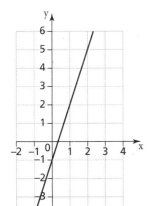 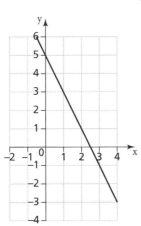 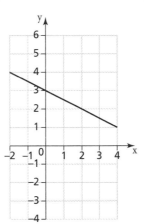 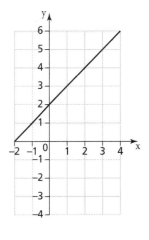

....

3 Write down the gradient of each line.

a) $y = 3x – 2$ ....

b) $y = –2x + 1$ ....

c) $y = x + 4$ ....

d) $y = 8 – 3x$ ....

e) $y = 2x – 9$ ....

f) $y = 9 – 7x$ ....

4 Is each statement about the graph of $y = 4x + 3$ true or false?

	True	False
It is a straight-line graph.	☐	☐
It passes through the point (−2, 5).	☐	☐
It passes through the point (3, 15).	☐	☐

5 Write each equation in the correct position in the table.

$y = 6$

$y = 3x^2$

$y = 7 - 2x$

$xy = 4$

$y = 11 - x^2$

$y = 7x + 6$

Equations that correspond to straight-line graphs	Equations that do not correspond to straight-line graphs

6 Pierre completes this table of values.

x	−1	0	1	2
y	−4	−1	4	5

He says, 'My table of values is for a straight-line graph.'

Without drawing a graph, how can you tell that Pierre has made a mistake?

....

....

7 Match the equations to the graphs.

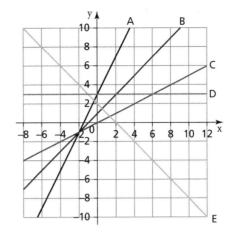

Equation	Graph
$y = 2x + 3$	
$y = x + 1$	
$y = \frac{1}{2}x$	
$y = -x + 2$	
$y = 3$	

8 Amy thinks that the graphs drawn on the grid have equations $y = 4x + 4$ and $y = 4x - 4$.

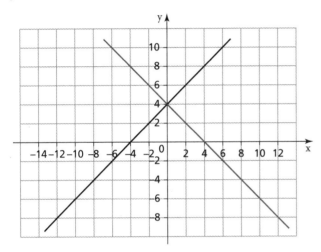

Explain how you can tell that she is wrong.

....

....

9 a) Write the equations of three lines that are parallel to the line $y = 5 + 2x$.

....

b) Write down the equations of three lines that go through the point $(0, -2)$.

....

Summary of key points

The amount a painter charges (c) depends on the area (a m²) he has to paint.

He uses this formula to calculate the charge:

$$c = 4a + 20$$

This relationship is shown on the graph.

The gradient of the graph is $\frac{120}{30} = 4$.

This shows that the total amount the painter charges increases by $4 for each square metre he paints.

The intercept on the vertical axis is $20. This is the fixed charge.

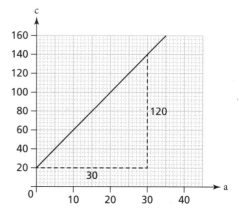

Exercise 3

1 Birthday balloons are sold online.

The price of a balloon is $2 each. The postage charge is $2.50 per order.

a) Complete the table for the price of buying different numbers of birthday balloons.

x Number of balloons	1	2	3	4	5	6	7	8
y Cost of balloons ($)								

b) Draw a graph to show the relationship.

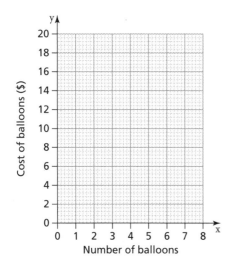

2 The graph shows the cost *C* ($) of ordering *n* bags of soil from a company.

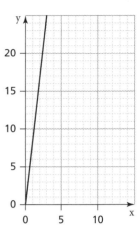

Calculate the gradient of the line and explain what it represents.

....

....

3 The cost, $C, or hiring a digger for *d* days is given by the function $C = 12d + 20$.

Tick the two statements that are true.

A The fixed cost of hiring a digger is $12. ☐

B The daily charge is $12. ☐

C The daily charge is $20. ☐

D The fixed cost of hiring a digger is $20. ☐

4 Alice begins saving money on her 13th birthday.

Her grandmother gave her $20 for her birthday.

After that she started saving $2 each week.

a) Write a formula for the total amount of money ($M) that she has saved w weeks after her 13th birthday.

M = ...

b) Draw a graph to show the relationship.

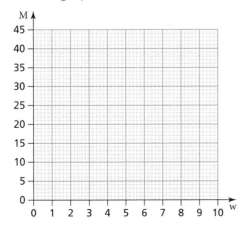

5 Amol buys a car. The value (y) of Amol's car after t years is given by the formula:

$y = 15\,000 - 1500t.$

a) Draw a graph showing the value of Amol's car.

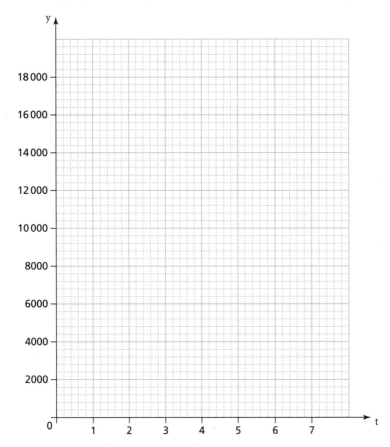

b) Amol thinks that the car cost $13 500 when new.

Explain why Amol is wrong. What should the answer be?

....

....

....

c) By how much does the value of the car decrease each year?

$...........

Think about

6 Why is the graph in question 5 not appropriate for values of t greater than 10?

Summary of key points

A **real-life graph** shows how something changes over time.

A **travel graph** shows how distance changes over time.

Example: Amy and Channa make the same journey between two towns.
The travel graph shows their journeys.

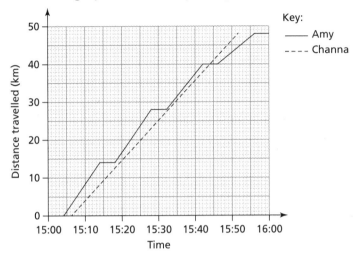

a) At what time does Amy first stop?

She stops when her graph is horizontal. This first happens at 15:14.

b) How far has Channa travelled when he passes Amy?

He passes Amy when their lines cross over. This happens when the distance
travelled is 40 km.

Exercise 4

1 **Hazlina and Abdul are brother and sister.**

**The travel graph shows their journeys to
school.**

a) **How many minutes does Hazlina's journey
to school take?**

.... minutes

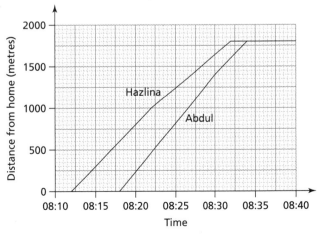

b) How many minutes after Hazlina left did Abdul leave for school?

.... minutes

c) How far from her home was Hazlina at 08:20?

.... metres

d) Work out the difference between Hazlina's and Abdul's journey times to school.

.... minutes

e) How far from school was Abdul at the time Hazlina arrived at school?

.... metres

2 The travel graph shows journeys made by two trains.

Train A travels from Central Station to North Station.

Train B travels from North Station to Central Station.

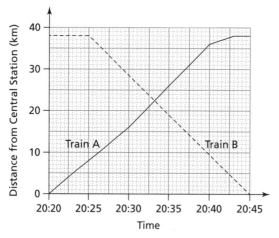

a) What is the distance between the two stations?   .... km

b) How long after Train A left Central Station did Train B leave North Station?

.... minutes

c) i) At what time did the trains pass each other?   ....

ii) How far from Central Station were the two trains at this time?   .... km

3 Nina and Jonas each walk the same route from a restaurant to a theatre.

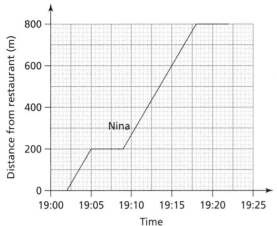

a) At what time did Nina arrive at the theatre?   ………..

b) Nina stopped at a shop on the way to the theatre. How long did Nina spend in the shop?

                                                  ……….. minutes

Jonas left the restaurant 5 minutes after Nina. He walked straight to the theatre. His journey took exactly 10 minutes.

c) Jonas says that the graph for his journey does not have a horizontal part.

   Is he correct? Explain why.

   ………………….…………………………….…………………….……………..…….

   ………………….…………………………….…………………….……………..…….

d) Show Jonas' journey on the travel graph.

e) At what time did Jonas pass Nina?   ………..

4 Helen buys a laptop.

The value (y) of the laptop after t years is given by the formula $y = 700 - 60t$.

Ben buys a different computer.
He pays $840. The value decreases by $95 each year.

By drawing a graph for Helen and Ben, find out after how long both computers will have the same value.

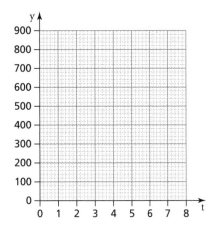

5 Leonora and Pasindu each have tanks that can hold 200 litres of water.

Leonora's tank is empty at the start. It takes her 8 minutes to fill her tank.

Pasindu's tank is one quarter full at the start. It takes Pasindu 10 minutes to fill his tank.

Leonora says that the containers will only have the same amount of water when they are full.

Use graphs to check whether Leonora is correct.

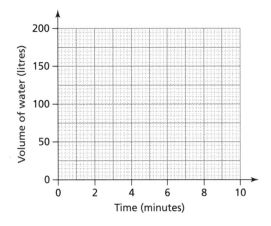

Thinking statistically

You will practice how to:

- Record, organise and represent categorical, discrete and continuous data. Choose and explain which representation to use in a given situation:
  - o Venn and Carroll diagrams
  - o tally charts, frequency tables and two-way tables
  - o dual and compound bar charts
  - o pie charts
  - o frequency diagrams for continuous data
  - o line graphs and time series graphs
  - o scatter graphs
  - o stem-and-leaf diagrams
  - o infographics.
- Use knowledge of mode, median, mean and range to compare two distributions, considering the interrelationship between centrality and spread.
- Interpret data, identifying patterns, trends and relationships, within and between data sets, to answer statistical questions. Discuss conclusions, considering the sources of variation, including sampling, and check predictions.

24.1 Comparing and using graphs

Summary of key points

You can compare the information given in two statistical graphs.

Example: The pie charts show the favourite fruits of the students and teachers in a school. Compare the information given in the pie charts.

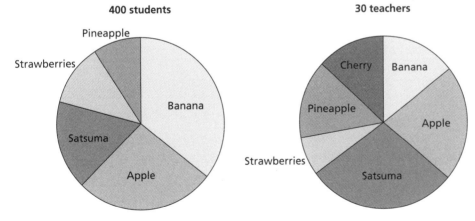

For example, the pie charts show:

- apples were the favourite fruit for a smaller proportion of teachers than students
- bananas were the most popular fruit among the students but satsumas were the most popular fruit among the teachers.

The pie charts do not show that a larger number of teachers chose satsuma as their favourite fruit as the total number of teachers is much smaller than the total number of students.

1 The pie charts show how Max and Nadia spend their monthly pay.

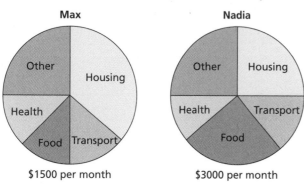

Max

Nadia

$1500 per month

$3000 per month

Tick to show if these statements are true or false.

	True	False
Max spends a greater proportion of his monthly pay on housing than Nadia.	☐	☐
Nadia spends a smaller proportion of her monthly pay on food than Max.	☐	☐
Max spends more money per month on health than Nadia.	☐	☐
Nadia uses more money per month for other things than Max.	☐	☐

2 The pie charts show the types of vehicle using a bridge on two different mornings.

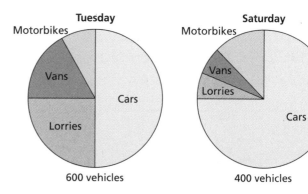

Tuesday

Saturday

600 vehicles

400 vehicles

a) Compare the proportion of the vehicles that are lorries using the bridge on the two days.

....

....

b) Compare the proportion of the vehicles using the bridge on the two days that are cars.

....

....

c) Compare the number of cars that use the bridge on each of the two days.

....

....

3 The pie charts show the types of tree in two woodlands.

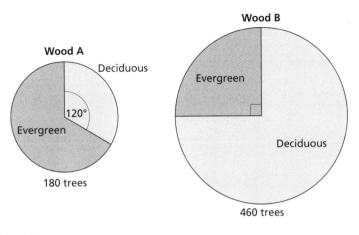

Evergreen trees keep their leaves in winter. Deciduous trees lose their leaves in winter.

a) Give a reason why the pie chart for Wood B might have been drawn with a larger radius than the pie chart for Wood A.

....

....

b) Compare the proportions of evergreen trees in the two woods.

....

....

c) Which wood has the larger number of evergreen trees?

Wood A ☐   Wood B ☐

Show how you worked out your answer.

....

....

4 The pie charts show the crops grown in two farms.

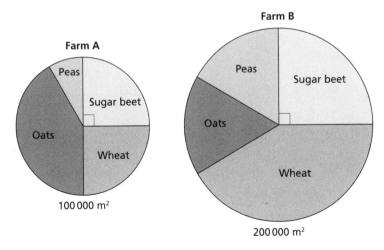

a) Draw a ring around the crop that covers the most land in Farm A.

peas   oats   sugar beet   wheat

b) Compare the proportion of land that is used to grow wheat in each of the two farms.

....

....

c) Explain how you can tell that a larger area of land is planted with sugar beat on Farm B than Farm A.

....

....

5 **A magazine runs a photography competition. People enter the competition by sending in a photograph of either a plant or an animal. A judge awards a score to each photograph that is entered.**

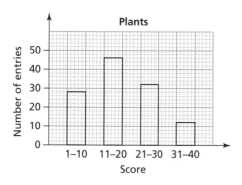
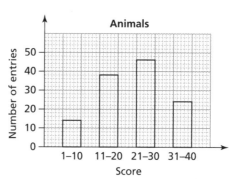

a) Compare the scores given by the judge to photographs of plants with the scores given to photographs of animals.

....

....

b) How many more photographs of animals were entered than photographs of plants?

....

c) Entries that were given a score of 31 or more were awarded a certificate. The organisers claim that 15% of all the entries were awarded a certificate.

Are they correct? Show your working.

....

....

....

6 A teacher gives her class a multiplication test. She compares the marks with the results the class got in the same test 3 months ago.

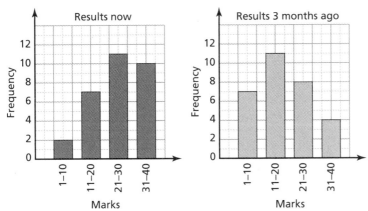

a) Which interval is the modal class for marks in the test now? ....

b) What is the modal class for marks in the test taken 3 months ago?

....

c) How do the results of the class now compare with the results 3 months ago?

....

....

7 All the students in one year group at a school take a maths quiz.
The frequency diagram shows their scores.

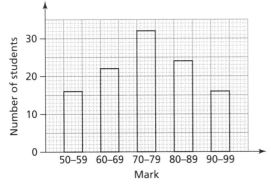

Certificates are presented to students that score 70 marks or more.
The pie chart shows the types of certificate these students receive.

The headteacher says that 15 students receive a gold certificate.

Is the headteacher correct? Show how you work out your answer.

....

....

....

....

8 A school puts on a play.
The bar chart shows the number of adults and children in the audience on each of the three nights.

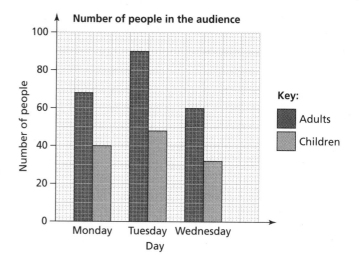

On which night was there the greatest proportion of children in the audience?

....

Show how you worked out your answer.

....

....

....

24.2 Comparing distributions using measures of average and spread

Summary of key points

The mean, median and mode are measures of **average**. They give a typical value for a set of data. They are useful when deciding whether the values in one set of data are larger or smaller overall than values in a second set of data.

The range is a measure of **spread**. It is used to compare the variation in the values in sets of data. Values in one data set are more consistent than in another if the range is smaller.

Exercise 2

1 Bridgette and Erika are competitive swimmers. The table summarises their times in their last 10 races.

	Median time (seconds)	Range of times (seconds)
Bridgette	72.3	3.9
Erika	71.6	7.2

Use the correct names to complete these statements.

.... is faster on average than ....

....................'s swimming times are more consistent than ....'s.

2 Teams from two schools take part in a running tournament.
The times for the runners in each team are summarised below.

School A

Mean time: 47 minutes

Range of times: 11 minutes

School B

Mean time: 45 minutes

Range of times: 17 minutes

a) Which team was faster on average?

School ....

Explain how you know. ....

....

b) Which school had times that were more consistent?

School ....

Explain how you know. ....

....

3 A class takes tests in Science and Maths.

Both tests are marked out of 50.

Science test

Median mark: 33

Range of marks: 17

Maths test

Median mark: 30

Range of marks: 23

a) Anton says, "The marks were higher on average in the Maths test than the Science test."

Anton is **not** correct. Explain why. ....

....

b) Anton says, "The marks in the Maths test were more variable than the marks in the Science test."

Is Anton correct?   He is correct. ☐   He is not correct. ☐

Explain your answer. ....

....

4 Two classes of 30 children take a test in Maths. Some information about the results is given below.

Class 8A	Class 8B
Mean mark: 65	Mean mark: 55
Median mark: 59	Median mark: 56
Range of marks: 72	Range of marks: 48

Tick to show if these comparisons are true or false.

	True	False
The marks for class 8A are more spread out than the marks for class 8B.	☐	☐
The marks for class 8A are higher on average than the marks for class 8B.	☐	☐
Less than half of the children in class 8B scored more than 50 marks.	☐	☐
The total of all 30 marks for class 8A is greater than the total of all 30 marks for class 8B.	☐	☐

5 The stem-and-leaf diagram shows the number of visitors to an exhibition on 20 Mondays.

```
20 | 5  6  8
21 | 0  1  5  5  8
22 | 1  3  5  6  7  8  9
23 | 3  7  8
24 | 1  4
```

| Key 20 | 5 represents 205 visitors |

a) Find the median number of visitors. ....

b) Write down the modal number of visitors. ....

c) Calculate the range for the number of visitors. ....

The table shows some information about the number of visitors to the same exhibition on 20 **Tuesdays**.

Median	239 visitors
Mode	229 visitors
Range	79 visitors

d) Compare the number of visitors on Mondays and Tuesdays.

....

....

....

....

25 Accurate drawing

You will practice how to:

- Construct triangles, midpoint and perpendicular bisector of a line segment, and the bisector of an angle.
- Represent front, side and top view of 3D shapes to scale.

25.1 Construction of triangles

Summary of key points

Given two angles and a side (ASA):

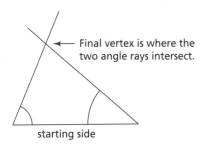

Final vertex is where the two angle rays intersect.

starting side

Given two sides and an angle (SAS):

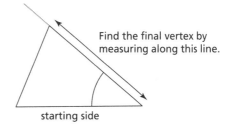

Find the final vertex by measuring along this line.

starting side

Given three sides (SSS):

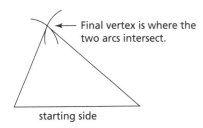

Final vertex is where the two arcs intersect.

starting side

Given a right angle, the hypotenuse and a side (RHS):

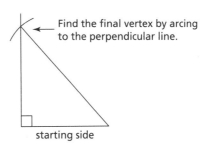

Find the final vertex by arcing to the perpendicular line.

starting side

Exercise 1

1 **The diagram shows a sketch of triangle *ABC*.**

   a) Draw the triangle accurately.

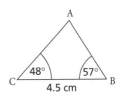

b) Measure the length AB.

.... cm

2 **The diagram shows a sketch of triangle *DEF*.**

a) Draw the triangle accurately.

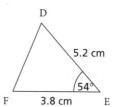

b) Measure the angle DFE.

.... °

3 In triangle *PQR*:

QR = 50 mm   angle *PQR* = 120°   angle *PRQ* = 25°

a) Draw the triangle PQR accurately.

b) Measure the length PR. Give your answer in millimetres.

.... mm

4 Construct a triangle with sides measuring 5.2 cm, 3.7 cm and 4.5 cm.

5 Construct a triangle *ABC* with angle *ABC* = 90°, *AC* = 5 cm and *AB* = 3.5 cm.

6 Explain why it is not possible to draw a triangle with sides measuring 6 cm, 3 cm and 2 cm.

....

....

....

Summary of key points

Constructing the perpendicular bisector
of line segment AB:

Constructing the bisector of angle ABC:

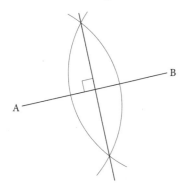

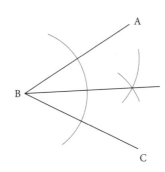

Exercise 2

1 **Draw the perpendicular bisector of each line segment.**

a)

b)

A————————B

2 Construct the bisector of each angle.

a)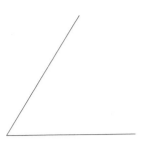

b)

3 Solomon tries to bisect angle *ABC*.

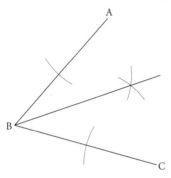

Explain the mistake Solomon has made.

....

....

4 The diagram shows line segments *AB* and *BC*.

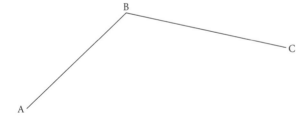

a) Construct the perpendicular bisector of line segment AB.
   Label the midpoint, M, of AB.

b) Construct the perpendicular bisector of line segment BC.
   Label the midpoint, N, of BC.

c) Find the midpoint of line segment MN using compasses.
   Label this point P.

Summary of key points

A 3D shape can be drawn to scale on paper using plans and elevations.

The **plan view** is the view of the shape from the top.

An **elevation** is drawn from the side or from the front.

Exercise 3

1. Draw on the centimetre grid, the plan, the side elevation and the front elevation of the shape shown below. Use the scale 1 cm to 1 m.

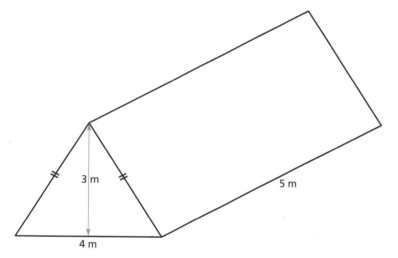

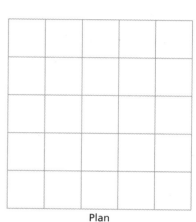

Plan

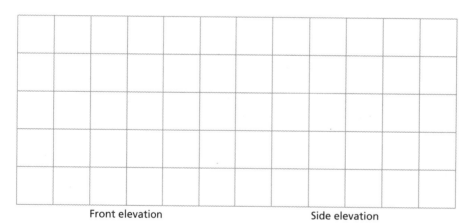

Front elevation   Side elevation

2 Draw on the centimetre grid, the plan, the side elevation and the front elevation of the building shown below. Use the scale 1 cm to 2 m.

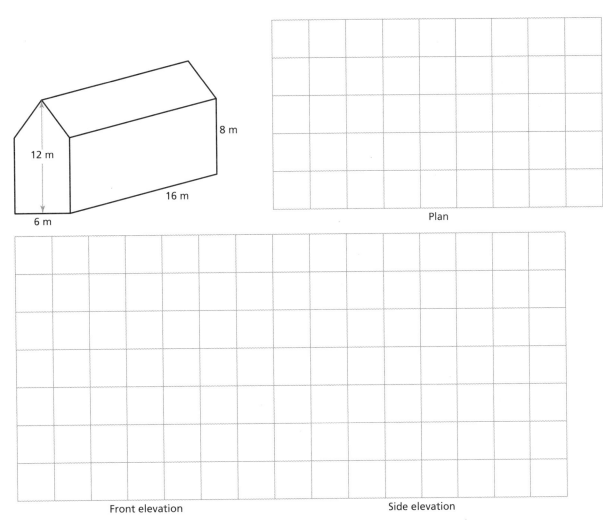

Plan

Front elevation   Side elevation

3 Draw on the centimetre grid, the plan, the side elevation and the front elevation of the block shown below. Use the scale 1 : 200.

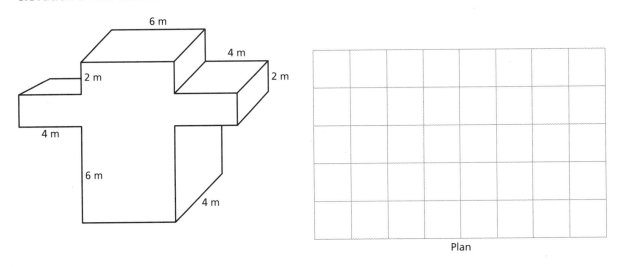

Plan

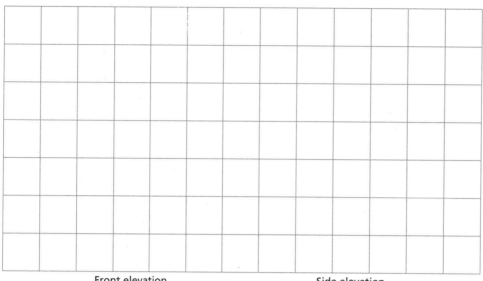

Front elevation   Side elevation

4 Michael drew the plan and elevations of a structure.

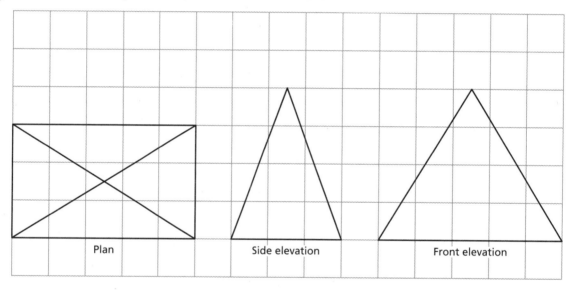

Plan   Side elevation   Front elevation

He drew them on centimetre grid paper and to a scale of 1 : 300.

Sketch the structure Michael drew the plan and elevation for, and describe it as fully as possible.

....

....

....

....

....

....